Quality of Life in Asia

Volume 13

This series, the first of its kind, examines both the objective and subjective dimensions of life quality in Asia, especially East Asia. It unravels and compares the contours, dynamics and patterns of building nations by offering innovative works that discuss basic and applied research and emphasizing inter- and multi-disciplinary approaches to the various domains of life quality. The series appeals to a variety of fields in humanities, social sciences and other professional disciplines. Asia is the largest, most populous continent on Earth, and it is home to the world's most dynamic region, East Asia. In the past three decades, East Asia has been the most successful region in the world in expanding its economies and integrating them into the global economy, offering lessons on how poor countries, even with limited natural resources, can achieve rapid economic development. Yet while scholars and policymakers have focused on why East Asia has prospered, little has been written on how its economic expansion has affected the quality of life of its citizens. This series publish several volumes a year, either single or multiple-authored monographs or collections of essays.

More information about this series at http://www.springer.com/series/8416

Ming-Chang Tsai · Noriko Iwai
Editors

Quality of Life in Japan

Contemporary Perspectives on Happiness

Springer

Editors
Ming-Chang Tsai
Academia Sinica
Taipei, Taiwan

Noriko Iwai
Osaka University of Commerce
Osaka, Japan

ISSN 2211-0550 ISSN 2211-0569 (electronic)
Quality of Life in Asia
ISBN 978-981-13-8909-2 ISBN 978-981-13-8910-8 (eBook)
https://doi.org/10.1007/978-981-13-8910-8

This Springer imprint is published by the registered company Springer Nature Singapore Pte Ltd.
The registered company address is: 152 Beach Road, #21-01/04 Gateway East, Singapore 189721, Singapore

Contents

Editors and Contributors

About the Editors

Ming-Chang Tsai is Research Fellow and Deputy Director of Research Center for Humanities and Social Sciences, CHSS, Academia Sinica, Taiwan. He is President of the International Society for Quality of Life Studies and was former President of Taiwanese Sociological Association and President of Research Committee of Social Indicators (RC55) of the International Sociological Association. His current research project focuses on family relationship, youth, and aging in comparative perspective. His articles have appeared in *Comparative Sociology*, *Globalizations*, *Journal of Marriage and Family*, *Journal of Contemporary Asia*, and *Sociological Research Online*. His recent books include *Global Exposure in East Asia* (Routledge, 2015) and *Family, Work and Wellbeing in Asia* (co-editor, Springer, 2017) (mtsai304@gate.sinica.edu.tw).

Noriko Iwai is Director of the Japanese General Social Survey Research Center and Professor of Faculty of Business Administration, Osaka University of Commerce. She is a principal investigator of JGSS and East Asian Social Survey, an Executive Director of the Japan Sociological Society, and a member of Science Council of Japan. Her current research project is supporting Japanese researchers in the humanities and social sciences to prepare their data for public usage. She has written on family and on trends in Japanese people's opinions. Her recent articles are "Division of housework in Japan, South Korea, China and Taiwan" in *Family, Work and Wellbeing in Asia* (Springer) and "The Impact of the Great East Japan Earthquake and Fukushima Daiichi Nuclear Accident on People's Perception of Disaster Risks and Attitudes Toward Nuclear Energy Policy" in *Asian Journal for Public Opinion Research* (n-iwai@tcn.zaq.ne.jp).

Contributors

Dang Thi Viet Phuong Vietnam Academy of Social Sciences, Hanoi, Vietnam

I. Hamada Monash University, Melbourne, Australia

Hiroo Harada Senshu University, Tokyo, Japan

Hachiro Iwai Kyoto University, Kyoto, Japan

Noriko Iwai Japanese General Social Survey Research Center, Osaka University of Commerce, Osaka, Japan

Masayuki Kanai Senshu University, Tokyo, Japan

Xinxin Ma Center for Far Eastern Studies, University of Toyama, Toyama, Japan

N. Oishi Monash University, Melbourne, Australia

Xiangdan Piao Departments of Urban and Environmental Engineering, School of Engineering, Kyushu University, Kyushu, Japan

Takayuki Sasaki JGSS Research Center, Osaka University of Commerce, Higashi-osaka, Japan

Katsumi Shimane Senshu University, Tokyo, Japan

Kuniaki Shishido JGSS Research Center, Osaka University of Commerce, Higashi-osaka, Japan

Eiji Sumi Niigata University, Niigata, Japan

Ming-Chang Tsai Research Center for Humanities and Social Sciences, Academia Sinica, Taipei, Taiwan

Introduction

An Introduction to Quality of Life in Japan: Contemporary Approaches

Ming-Chang Tsai and Noriko Iwai

Abstract This chapter gives a brief overview of current approaches to studying quality of life in Japan. After discussing observations from mass media and anthropological research, an empirical approach to subjective well-being for Japanese society is proposed in order to better understand the levels and patterns of life satisfaction in various domains among Japanese males, females and emigrants. The main findings of each paper in this book are also highlighted.

Keywords Quality of life · Gender difference · Aging population · Income effect · Immigration

Japan has been one of the most envied societies in the contemporary era. It has outperformed many other wealthy countries in major indicators of economic growth, democratization, and social development. According to the Human Development Index proposed by the United Nations, as of 2017, Japan ranked 19th among 189 countries in the world. The HDI is a composite index that incorporates life expectancy, educational opportunities and national income level (UNDP 2018). Wealthy Western countries that have been considered examples of high quality of life, such as Luxembourg, France, Spain or Italy, in fact lag behind Japan on this widely used index of living conditions. Also, Japan ranked 22nd among 167 countries in 2018 in the Democracy Index compiled by the Economist Intelligence Unit.[1]

A general indicator like the HDI may miss the detailed patterns or disparity of well-being in a society. Japan has been reported as a typical, strongly patriarchal society in which males dominate major social institutions such as family, enterprises, and government (Sechiyama 2013). It is worth paying adequate attention to this aspect. In the often-cited Gender Inequality Index (GII), another indicator in the

[1]The Democracy Index 2018. https://www.eiu.com/topic/democracy-index.

M.-C. Tsai (✉)
Research Center for Humanities and Social Sciences, Academia Sinica, Taipei, Taiwan
e-mail: mtsai304@gate.sinica.edu.tw

N. Iwai
Japanese General Social Survey Research Center, Osaka University of Commerce, Osaka, Japan
e-mail: n-iwai@tcn.zaq.ne.jp

M.-C. Tsai and N. Iwai (eds.), *Quality of Life in Japan*, Quality of Life in Asia 13,
https://doi.org/10.1007/978-981-13-8910-8_1

UN's Human Development Project, Japan ranked 22nd among 160 countries. GII uses three dimensions: reproductive health (measured by the maternal mortality ratio and the adolescent fertility rate), empowerment (women's share of legislative seats as well as of secondary education) and labor market participation. As of 2017, Japan holds a position among the top tier, with a rank of 22nd, and fares better than the US, UK or New Zealand (UNDP 2018).

In a recent report on the Mother's Well-being index, which ranks 150 countries in the world in 2010, Japan ranked 29th (Tsai and Tai 2018). This index considers not merely maternal health differences, but also policy support for mothers with regard to their entitlements to maternal leave and reentering the labor market. Because this measure concerns welfare policy inputs, higher-income countries, unsurprisingly, occupy most of the top positions. Japan does not perform as well as it does on the HDI, but it is still in the first tier in support for mothers' well-being. Thus, policies supporting mothers have been brought in, and the percentage of females who continued to work after childbirth has increased. However, it still remains at a low 53.1% for those who gave birth in 2010–2014.[2]

There is another index which has been frequently referred to: the Global Gender Gap Index, reported by the World Economic Forum. Japan does not seem to fare so well here: it ranked 110th among 149 countries in 2018.[3] GGI is a composite of four sub-indices. Japan has smaller gender gaps in health and survival[4] (41st) and educational attainment[5] (65th), but larger gender gaps in economic participation and opportunity[6] (117th) and in political empowerment (125th). When it comes to economic inequality between genders, in contrast to health or educational attainment, Japan shows obvious weakness. The Gender Equality Bureau of the Cabinet Office, Japan, posts the above gender equality indices on its website[7] and has proposed measures to substantially reduce the gender gaps in opportunities, benefits and responsibilities.[8]

In contrast to the relatively highly regarded performance on economic growth, democratization, and social development, and to the reserved evaluation on gender equality, some recent observations on Japan cast a totally different picture. Public media have painted a bleak picture of worsening life conditions in Japan which feature fears, worries and discontentment, due to "the increasing cases of workplace-

---

[2]See *Annual Health, Labour and Welfare Report 2017*, p. 181. The original source of the data is "Fifteen Japanese National Fertility Survey (Survey on Married Couples)", National Institute of Population and Social Security Research.

[3]http://www3.weforum.org/docs/WEF_GGGR_2018.pdf.

[4]The sex ratio at birth is 0.95, which ranked 1st. Women live a healthy life four years more than men (57th).

[5]Other than the ratio of enrollment in tertiary education (0.95), there is no gender difference in literacy, or enrollment in primary and secondary education.

[6]The ratio of female legislators, senior officials and managers to males is 15% (129th), which is less than half of the world average, 33%. The ratio of women to men in the legislature is 11% (ranked 130th), which is again less than half of the world average, 28%.

[7]http://www.gender.go.jp/international/int_syogaikoku/int_shihyo/index.html.

[8]http://www.gender.go.jp/english_contents/index.html.

induced depression, number of suicides and deaths through overwork (*karoshi*), the mounting fear of solitary death (*kodokushi*), and people withdrawing from society (*hikikomori*)" (Manzenreiter and Holthus 2017: 1). Many stories gathered from mass media seem to represent "a society hampered by maladaptation" in a post-growth phase in which "increasing proportions of its members, across all age groups, are threatened by dissatisfaction, deprivation, alienation, depression, fear, and hopelessness" (Manzenreiter and Holthus 2017: 1). These and other stories of misery as well as psychopathologies reported in popular media indicate that Japan is not a promising research site for studies on happiness. Indeed, the official statistics of Japan report some similar trends about *karoshi*,[9] *kodokushi*[10] *or hikikomori*.[11]

However, anthropologists of happiness in Japan have argued we should avoid a one-dimensional understanding as such, and suggest a panoptical perspective to look at the society in its entirety. They particularly propose examination of mundane, ordinary practical expression of happiness, and distinctive structural or institutional contexts in which (un)favorable emotions originate or are embedded. Anthropological approaches usually direct research interest into diverse sites where ordinary people are observed giving sensory affect, personal interpretations, cognitive judgements, and concluding remarks on their daily life. As Manzenreiter and Holthus (2017: 2) indicate, anthropological approaches are particularly interested in "explorations of the variety of notions and expressions of happiness, which are conceived of so differently in distinct social and cultural contexts that some researchers are prompted to ask if there remain any commonalities".

Even when happiness is observed, anthropological interest lies often in "what anyone means by happiness" (Colson 2010). Happiness, essentially an internal state, is only captured partially from externalities. Diverse measures of all ranges of emo-

---

[9]According to a 2018 white paper on promotion of measures for karoshi (https://www.mhlw.go.jp/wp/hakusyo/karoushi/18/dl/18-1.pdf) and the 13th Occupational Safety and Health Program (https://www.mhlw.go.jp/content/11200000/000341159.pdf) by the Ministry of Health, Labour and Welfare, the number of requests for workers' accident compensation due to cerebrovascular/heart diseases caused by an overload of work has fluctuated around 830 cases after its peak year, 2006 (938 cases). However, the number of requests due to mental health disorders caused by a significant psychological burden increased from 212 cases in 2000 to 1,932 cases in 2017. Approximately 60% of the workers were affected by severe anxiety, distress and stress relating to their job or occupational lives in 2002 and in 2016. However, note an opposite trend in work-related suicides: the number of suicides induced by job-related issues such as fatigue, unhappy relationships with coworkers, failure on the job or changes in work environment declined during the same period (2,207 cases in 2007 and 1991 cases in 2017). The total number of suicides similarly decreased, from 33,093 cases in 2007 to 21,321 cases in 2017 according to the above government report.

[10]According to a white paper on aging society by the Cabinet Office, the number of solitary deaths of people over 65 in Tokyo's 23 wards has increased from 1,451 in 2003 to 3,179 in 2016 (https://www8.cao.go.jp/kourei/whitepaper/w-2018/html/zenbun/s1_2_4.html).

[11]The number of people withdrawing from society is estimated to have expanded in Japan based on surveys on life among young people and middle aged, while the estimated population withdrawing from society among people aged 15 to 39 decreased from 696 thousand in 2010 to 541 thousand in 2016 (https://www8.cao.go.jp/youth/kenkyu/hikikomori/pdf/kosshi.pdf), it was estimated that 613 thousands of people withdrawing from society among people aged 40 to 64 in 2018 (https://www8.cao.go.jp/youth/kenkyu/life/h30/pdf/kekka_gaiyo.pdf).

tions are thus needed to grasp the different states of being contented, joyous, glad, lucky, fortunate, jubilant, and so on. How these emotions are particularly apparent at certain moments is equally intriguing. Specific times and spectacular events in which people feel elated are given particular attention. St. John (2012) proposes to look at festivals, in which major common sources of human happiness concentrate on site: chanting, dancing, laughing, social clubbing, and collective trance. Carnivals and competitive sport (particularly between "rival" cities or countries) can also belong to this group. In all these social gatherings, there is a religious element, which has been called the realm of "collective effervescence" (St. John 2012: 10). This way of observing happiness necessarily concludes that a festive event, given its high pleasure, tends to be radical, short-termed, indeterminate, uncertain, and transient. To employ a Durkheimian term, eventful happiness perceived as such constitutes a fundamental "social category" to be differentiated from dry, dull daily routines—there is a clear boundary between the two worlds. Yet in modern society, especially in East Asian societies in which emotional moderation rather than affective catharsis is regarded as more proper in managing the human-environment relationship, a cult of ecstasy for some can be frowned upon or even punished by the social community or legal institution. An intentional pursuit of the "greatest happiness" is not a golden rule. But this does not defy the importance of well-being in human life. As the normative perspective has argued, happiness might not be the main driver generating what is good for individuals or their society, but no social institutional building or individual action can possibly justify itself without giving happiness a suitable role (Bartram 2012). It is critical to differentiate what sort of happiness or satisfaction is deemed desirable across cultures. It is on this issue that anthropological research on happiness seems to have contributed major insights.

The collected works of this book approach the phenomenon of happiness of Japanese by way of empirical investigation. This is a thoughtful response to the current efforts to understanding happiness in this society. Stories, anecdotes, and tales popularized in mass media are often exaggerations of specific cases or personas, and are unsuitable for evaluation of the mass public's life experiences and consequential well-being. Anthropology as a discipline has a strong merit in providing in-depth description of what is happening in situ, yet it is quite limited in translating specific observations into a general account of a society. In particular, the variations across subpopulations in their feeling of happiness are not examined systematically in a way allowing the levels and differences with regard to how satisfied people feel with lives to be measured, compared and accounted for. An empirical approach is a necessary component of a cross-disciplinary literature to better understand the well-being of the Japanese population. Unfortunately, research outcomes from this approach have been less available. It is this paucity that this book attempts to remedy by providing analysis of various aspects of subjective well-being and specification of the potential factors underlying happiness among Japanese males and females. Working on these issues, authors of this book extensively utilize survey data collected from large, national samples to ensure representativeness. This way of examining the patterns of happiness and evaluating their potential influences provides an effective pathway to answering very fundamental questions about the quality of life in Japan.

1 Does Japan Fare Less Well in Quality of Life?

Japan does not perform less well in terms of the individual level of well-being from a cross-national comparative perspective, despite frustrating observations from mass media. In Fig. 1, life satisfaction of Japanese is assessed along with three other East Asian societies, as well as the US and Germany, by a very common measure which solicits a response to this question, "Where would you put your satisfaction with your life as a whole?" (on a ten-point scale). We display the means of the six societies from the first wave to the most recent one.[12] Across a time period of twenty years, the US appears to score the highest, with Germany being second (included in the survey only starting from the third wave). Over time, Japan increased from a mean score of 6.6 to nearly 7.0, still lagging behind the two wealthy Western societies but surely showing an upward trend. China, Taiwan and South Korea registered lower scores and are placed in the lower part of the figure. While the recent surge of life satisfaction in Taiwan and China drew them near to that of Japan (in wave 6), South Korea is slower in catching up.

Figure 1 also reveals that national income plays a role in determining the cross-national difference in self-rated well-being. Happiness at the societal level appears to go up with increasing national income. The US is an exception, however. The US has been extensively studied as an example of what is called the Easterlin Paradox: it is likely that for wealthy countries, as national income escalates, the level of

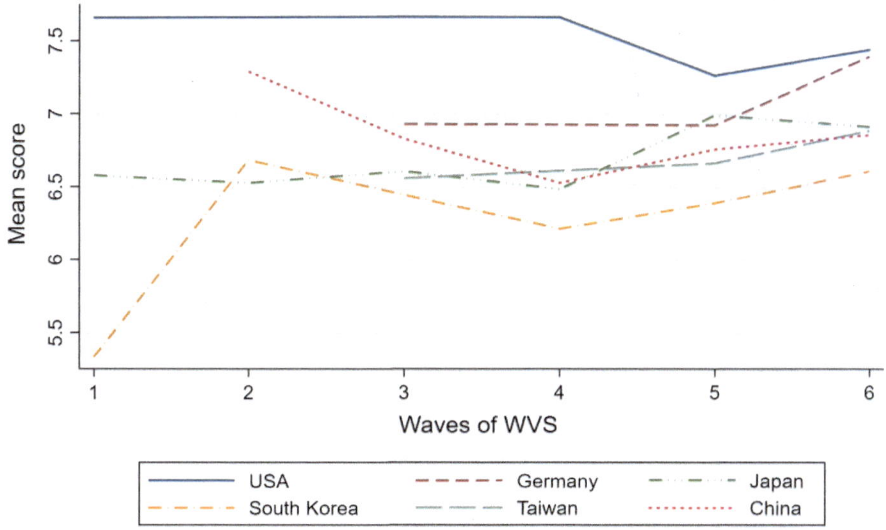

Fig. 1 Trends of life satisfaction in six countries (World Values Survey)

[12]TThe World Values Survey conducted a multiple country survey with an interval of every five years. The first wave was conducted during 1981–1984; wave 2, 1989–1993; wave 3, 1994–1999; wave 4, 1999–2004; wave 5, 2005–2008; and wave 6, 2010–2014.

perceived well-being does not follow its economic growth trend closely (Easterlin 2009; Easterlin et al. 2010). Japan is also mentioned as a case of this paradox, particularly referring to the period after the post-war economic miracle[13] which ended in 1973. Although navigating into an economic bubble between 1986 and 1991, Japan in fact significantly advanced in terms of GNI per capita: from $13,820 to $28,540.[14] However, the level of perceived well-being remained unchanged between wave 1 and wave 2 (Fig. 1). In contrast, for lower- and middle-income countries, growth of national income is a factor contributing to subjective well-being.

I note another observation from Fig. 1. The magnitude of increase in life satisfaction over time is small. South Korea experienced a big surge between wave 1 and wave 2, or more specifically, between 1982 and 1990 when the World Values Survey was conducted in this society. Interpretation of a drastic uptrend can be open to a wide range of viewpoints. Yet, fast economic growth during the 1980s should play a role here. Indeed, South Korea's gross national income per capita increased to $6,360 in 1990 from $2,040 in 1982,[15] a remarkable growth record by any measure.

Following this introductory chapter, Shishido and Sasaki's chapter in this book also documents a rising trend of subjective well-being in Japan, compiling longitudinal data from the Japanese General Social Survey. This evidence is presented in a sophisticated model of age-period-cohort analysis. This model has a strong advantage of differentiating the period effect from the otherwise confounding influences of the age and cohort factors. In very general terms, Japanese increasingly felt happy during 2000–2015, rather than revealing a collective dismal outlook about life conditions. Two groups show deviations in their perceived well-being, however. First, the authors highlight that those born in 1935–1939 had a significantly lower level of happiness, perhaps because they grew up in the chaos of the post-war era, lost their father in the war, and had less access to adequate education. In addition, younger cohorts who were born in the 1980s reported a lower level of happiness. As the authors contend, Japan's bubble economy and tight labor market seem to have had an unfavorable impact on these "juniors" working in various industries.

2 Income, Time Poverty, and Well-Being in Japan

Income perhaps is the factor which receives more attention than others in explaining to what extent an individual's happiness is determined. The income effect hypothesis has long been tested in empirical studies to show that earned income, a proxy of an individual's material conditions, effectively increases life satisfaction across Asian countries with different levels of economic development (Chen 2012; Tsai

[13]Measuring National Well-Being: Proposed Well-being Indicators released in 2011 by the Commission on Measuring Well-being, Japan, established in 2010 by the Cabinet Office. https://www5.cao.go.jp/keizai2/koufukudo/pdf/koufukudosian_english.pdf.

[14]https://www.macrotrends.net/countries/JPN/japan/gni-per-capita Access date: May 6, 2019.

[15]https://www.macrotrends.net/countries/KOR/south-korea/gni-per-capita.

2015). Recent literature concentrates more attention on the influence of relative income, which has long been identified for its influence on an individual's attitude and behavior of consumption in the tradition of reference group and social comparison theories (Newcomb et al. 1965). Researchers have proposed different approaches to measure relative income, for instance, by comparison with one's parent, neighbors, or coworkers (Firebaugh and Schroeder 2009; McBride 2001). Harada and Sumi's chapter in this book utilizes an index of relative income on a par with the municipal average income. Their finding indicates a positive influence of relative income; yet they also show that the magnitude of the effect of both absolute and relative income is not as substantial as expected. Ma and Piao's article in this book also tested relative income's effect. They adopt an imputation approach by a function of a husband's education, work status and years, and urban residence. This equivalent income is used as a reference point against the reported income of a husband. The ratio of the household income to this imputed income, however, is found to have no correlation with happiness measured on five-point scale. In contrast, absolute income generates a substantial positive influence. It seems that income in general has a favorable influence on well-being in Japan nowadays; but its measures can vary and might not operate equally well across different social groups.

Imputed income from a combination of select demographic characteristics might not effectively reflect a reference group. Since individuals have different focal groups for comparison, it would be more interesting to identify an ideal referent for each specific group, rather than proposing one for use across all subpopulations. Moreover, this imputed group can be thought of as a *positive* reference group—when an individual earns an income close to or higher than that of this group, then favorable outcomes follow (that is, happiness increases). On the other hand, a *negative* reference group (Newcomb et al. 1965: 109) also deserves incorporation into research designs. This is a reference point from which an individual prefers to maintain a distance or a gap. The greater the income difference, the more a person is expected to feel satisfied about his or her socio-economic status and life condition. Evidence for this hypothesis, however, seems to be lacking, at least in this region.

Higher income usually means more time spent on work for certain populations. Therefore, an income effect on happiness can be offset to a certain extent. Ma's chapter on time poverty offers an interesting test of this hypothesis. Time poverty, or lack of discretionary time, is measured by possession of time besides committed activity (working, commuting, caring activities, etc.) and necessary activities (sleeping, eating, etc.). While this definition seems to lead to a measure for lack of leisure time, it is motivated more to capture flexibility and availability of precious *private* time. The finding shows that for both single mothers as well married mothers, the household income does not generate any impact on discretionary time. A suitable indicator of income effect would be income of the working mother, which is, unfortunately, not explored in this study. What seems to interest us is that time poverty is a strong predictor of happiness and mental health. In this sense, time perhaps is more valuable than money when it comes to affecting subjective well-being.

3 Ageing and Familial Bonds

Ageing has been a major demographic feature of Japan. As of 2015, people aged 65 and above constitute 26.6% of the population. By 2040 this figure is forecast to increase to approximately 35.3%, which is among the highest in the world (National Institute of Population and Social Security Research 2019). It is in this context that numerous demographic studies and public policy debates have been happening. With the aged population increasing, and the working population shrinking, financial support and care for old parents necessarily become an urgent social problem. Hachiro Iwai's paper on old parents co-residing with children provides informative analysis of how such families fared in the past decades, a noteworthy approach to understanding the changing economic conditions of the retired parents living with their young adult children. A major observation is that along a trend of declining income and pension of the elderly, many of them chose to live with their children, the income of whom also has dwindled during Japan's economic downturns. Iwai argued that for lower-income households, this type of co-residence reveals more mutual dependence across generations. This reflects a sort of family-oriented welfare regime, in contrast to those observed in Western democracies where welfare and economic security are largely institutionalized through governmental policies.

While Iwai's chapter looks at how the non-working elderly fare financially, Kanai, Shimane and Phuong's concentrates on the association of ancestor worship as a family religion with a feeling of happiness in both Japan and Vietnam. This issue, perhaps unconventional at first sight, represents a continuity of patriarchal lineage. It is through establishing a firm connection between the living and their ancestors that an individual can feel he or she is fulfilling a familial role, and thus establish a secure self-identity and earn respect from family and community. Their quantitative evidence shows in Japan and Vietnam that frequency of conducting memorials for ancestors is positively associated with happiness. Also, expectation of being worshiped by descendants is correlated with happiness in a similar way. This study is based on cross-sectional data from national surveys. There is a difficulty in establishing a causal correlation between the variables of concern. Their findings are preliminary but can stimulate future research interest in the direction of religious forms of familial bonds in both cultures.

Kanai et al.'s paper addresses another important issue, which pertains to the "oldest old" people in Japan. As many members of this group do not have a burial place or do not expect to have a relative to take care of their tomb, many of them have started to look for new ways of burial. A "communal grave" discussion group has become increasingly popular nationwide, in which participants who usually have very thin attachment to family gather and discuss their burial plans in advance. And through this collective meeting, a relationship, which is called "burial-friendship", has emerged, with these otherwise strangers evolving into a mutual bond. Alongside this new practice is a more elaborate "cherry blossom burial". The burial place is under a *sakura* tree chosen before death. Those who will be buried in a neighboring plot get together to discuss various issues of management with "grave-friends". This practice generates a new social network for the very old people as a result.

4 Gender, Intimacy and Migration

Studies of quality of life have regularly surveyed married people's perceived well-being, often comparing the married with the unmarried group to gauge how a good marriage can enhance one's level of happiness. There is no dearth of evidence for this hypothesis that marriage facilitates perceived well-being and that unmarried people fare less well in various indicators of life satisfaction among Japanese (Qian and Sayer 2016; Ngoo et al. 2015). One major reason that married people enjoy better emotional well-being is that they possess a close relationship with their partners, among other reasons (Gustavson et al. 2016). Yet what facilitates the sought-after intimacy for couples is not researched much from a cross-cultural perspective. For the case of Japanese couples, what constitutes the major determinants of their intimacy? Tsai's paper in this book offers comparative evidence on how intimacy is affected by a number of potential factors across Japan, South Korea, Taiwan and China. In contrast to measuring intimacy psychologically as a feeling of closeness or attachment, Tsai argues that intimacy can also be understood as willingness to disclose oneself to one's spouse, thus emphasizing more the reciprocal exchange of care and concern in conceptual operationalization. More interestingly, a four-country national survey dataset was used for assessment of the influence of assortative marriage and equity factors for intimacy. The findings indicate that similarity in either education or age does not generate a substantial effect. A major decisive factor is equal sharing of decision making power, which is positively associated with intimacy in a consistent manner across the studied East Asian societies. Moreover, a husband's contribution to housework also helps enhance intimacy in Japan and Taiwan. This paper offers research findings from a cultural context which conventionally features a patriarchal relationship between the couple. Young people nowadays might seek to avoid unfavorable outcomes in marital life by choosing a partner with similar or equal status. This is especially so for females, as they increasingly obtain better education, actively participate in the labor market, and contribute substantially to family finances, frequently not less than their husbands do. The findings reveal ironically that equity between the couple is more important than a wife's labor market behaviors or a husband's income contribution when it comes to cultivating intimacy.

The majority of the chapters included in this book present empirical studies using data collected from Japan as well as other East Asian countries for comparison. Oishi and Hamad's chapter on emigration of Japanese skilled professionals to Australia is distinctive in that qualitative methods are applied. To explore the motivation, agency and actions of this particular group, a standard questionnaire might not obtain the best results, since it is the distinctive types of people, rather than the average ones, that draw research attention in this sort of exploratory study design. For this reason, their study is included to represent an interesting case in which well-being plays a special role in the decision to emigrate among a selective group of contemporary Japanese. Oishi and Hamada highlight pursuit of gender equality as a key motive for unmarried female specialists. It is more discrimination against women than the barriers in their careers that had pushed them to seek jobs outside Japan. However, after the Fukushima disaster of 2011, female immigrants pointed to potential natural and nuclear disasters as one major reason for exit. This concern with the impacts of environmental risks constitutes a form of "lifestyle migration" rather than sheer economic opportunities. Yet lifestyle migration would be an oversimplified explanation, as Oishi and Hamada also observe that among the skilled workers they interviewed, slow economic growth and the government responses with regard to pensions and other policy malpractices are also cited for explaining why many *families*, rather than *single* men and women, decided to leave Japan. For the families with minor children, a "safe environment" is equivalent to a better quality of life. However,, this observation is solely based on middle-class families of Japan. Lower-income people in this highly industrialized society might have a very different idea of what is a better life, an issue for which this edited book does not offer research results.

5 Quality of Life in Japan: Looking Forward

How is contemporary Japan doing in terms of quality of life? This is perhaps a question too big to answer for all the chapters in this book. But there are some hints from empirical findings presented here as a collectivity. Life does not seem to have fared worse over time in this wealthy Asian society. Longitudinal observation of how people perceived and rated their life conditions seems to suggest that Japanese on average are able to enjoy a desirable level of life satisfaction. Compared to wealthy Western countries, life satisfaction among Japanese somewhat lags behind. Other East Asian countries appear to be closing their gap with Japan, probably because their economic growth continues while that of Japan has become slow in the past decade or so.

But the editors do not wish to concentrate attention solely on large pictures of happiness of the Japanese. This book primarily aims to examine the detailed structure of life situations as perceived by Japanese people, rather than evaluating it with a broad, large concept, so we have made great effort in decomposing the whole population into different subgroups and examining how one may succeed in seeking a desired way of life while others fall short in terms of the feeling of happiness. Thus, the gender disparity of well-being necessarily has to be explored in depth. That is why papers dealing with the effect of income of employed females, time use for working mothers, or the heads of older families who delayed retirement in order to earn additional income, etc., are included in this book. Of course, these observations and findings are not able to entirely satisfy our curiosity about the different forms of disparity of well-being in Japanese society. Yet, the research outcomes presented here can be seen as one first move to understand the diverse patterns of life quality as provided from an empirical approach.

The editors of this book also wish to encourage more work on quality of life in Japan. It is through a look into the experiences and expressions about life conditions that we can understand this society better—and in an evidence-based way. Again, we also know of recent endeavors with a qualitative perspective. There have been few exchanges between the two approaches, which is unfortunate, especially because researchers interested in this topic ideally should be interdisciplinary in using theories and welcoming different research traditions and field techniques (large-scale surveys, observation in natural settings, focus interviews, etc.). We hope that readers of this book at this moment find our chapters insightful as well as helpful in crafting future research agendas, be they quantitative or qualitative.

References

Bartram, D. (2012). Elements of a sociological contribution to happiness. *Sociology Compass, 6,* 644–656.

Chen, W. (2012). How education enhances happiness: Comparison of mediating factors in four East Asian countries. *Social Indicators Research, 106,* 117–131.

Colson, E. (2010). Happiness. *American Anthropologist, 114,* 7–8.

Easterlin, R. (2009). Lost in translation: Life satisfaction on the road to capitalism. *Journal of Economic Behavior & Organization, 71,* 130–145.

Easterlin, R. A., McVey, L. A., Switek, M., Sawangfa, O., & Zweig, J. S. (2010). The happiness–income paradox revisited. *Proceedings of the National Academy of Sciences, 107*(52), 22463–22468.

Firebaugh, G., & Schroeder, M. B. (2009). Does your neighbor's income affect your happiness? *American Journal of Sociology, 115,* 805–831.

Gustavson, K., Røysamb, E., & Borren, I. (2016). Life satisfaction in close relationships: Findings from a longitudinal study. *Journal of Happiness Studies, 17,* 1293–1311.

Manzenreiter, W., & Holthus, B. (2017). Introduction: Happiness in Japan through the anthropological lens (pp. 1–21).

McBride, M. (2001). Relative-income effects on subjective well-being in the cross-section. *Journal of Economic Behavior & Organization, 45,* 251–278.

National Institute of Population and Social Security Research. (2019). *Population Projections for Japan (2016–2065): Summary*. Tokyo: NIPSSR.

Newcomb, T. M., Turner, R., & Converse, P. E. (1965). *Social psychology: The study of human interaction*. London: Routledge & Kegan Paul.

Ngoo, Y. T., Tey, N. P., & Tan, E. C. (2015). Determinants of life satisfaction in Asia. *Social Indicators Research, 124,* 141–156.

Qian, Y., & Sayer, L. C. (2016). Division of labor, gender ideology, and marital satisfaction in East Asia. *Journal of Marriage and Family, 78,* 383–400.

Sechiyama, K. (2013). *Patriarchy in East Asia: A comparative sociology of gender*. Leiden: Brill.

St. John, G. (2012). Altered together: Dance festivals and culture life. *American Anthropologist, 114,* 9–10.

Tsai, M.-C. (2015). Happiness and sociability in a nonrecursive model: The US and Taiwan compared. In F. Maggino (Ed.), *A life devoted to quality of life—Festschrift in honor of Alex C. Michalos* (pp. 297–314). The Netherlands: Springer.

Tsai, M.-C., & Tai, T.-O. (2018). How are mothers faring across the globe? Constructing a new Mothers' well-being index and assessing its validity. *Applied Research in Quality of Life, 3,* 647–670.

UNDP (United Nations Development Programme). (2018). *Human development indices and indicators 2018 statistical update*. New York: United Nations.

Ming-Chang Tsai is a Research Fellow and Deputy Director of Research Center for Humanities and Social Sciences, CHSS, Academia Sinica, Taiwan. He is President of the International Society for Quality of Life Studies, and was former president of Taiwanese Sociological Association and President of Research Committee of Social Indicators (RC55) of the International Sociological Association. His current research project focuses on family relationship, youth, and ageing in comparative perspective. His articles have appeared in *Comparative Sociology*, *Globalizations*, *Journal of Marriage and Family*, *Journal of Contemporary Asia*, *Sociological Research Online*. His recent books include *Global Exposure in East Asia* (Routledge, 2015) and *Family, Work and Wellbeing in Asia* (coeditor, Springer, 2017) (mtsai304@gate.sinica.edu.tw).

Noriko Iwai is Director of the Japanese General Social Survey Research Center and Professor of Faculty of Business Administration, Osaka University of Commerce. She is a principal investigator of JGSS and East Asian Social Survey, an Executive Director of the Japan Sociological Society, and a member of Science Council of Japan. Her current research project is supporting Japanese researchers in the humanities and social sciences to prepare their data for public usage. She has written on family and on trends in Japanese people's opinions. Her recent articles are "Division of housework in Japan, South Korea, China and Taiwan" in *Family, Work and Wellbeing in Asia* (Springer) and "The Impact of the Great East Japan Earthquake and Fukushima Daiichi Nuclear Accident on People's Perception of Disaster Risks and Attitudes Toward Nuclear Energy Policy" *Asian Journal for Public Opinion Research* (n-iwai@tcn.zaq.ne.jp).

Happiness in Japan: A Hierarchical Age-Period-Cohort Analysis Based on JGSS Cumulative Data 2000–2015

Kuniaki Shishido and Takayuki Sasaki

Abstract The purpose of this paper is to examine factors affecting the happiness of Japanese people based on the JGSS cumulative data 2000–2015. JGSS is a repeated cross-sectional survey conducted once a year or every other years. Analyzing the pooled data of JGSS makes it possible to examine the cohort effect and period effect. In addition, we can achieve more generalized findings by examining the effect of variables at the individual level while controlling for the effect of cohorts and periods. We employed a hierarchical Age-Period-Cohort Analysis because individuals are nested within two social contexts: period (time points of surveys) and cohort (birth cohorts with 5-year interval). The period and the cohort were set at the group level, age and other independent variables predicting happiness were set at the individual level. The results of our analysis revealed that (1) the age effect followed a U-shaped curve, (2) happiness declined in 2003, (3) happiness was low in 1935–1939 birth cohort, (4) original social stratum and social opportunities in the early period of life affected happiness over the life course, (5) subjective relative household income had a stronger relation with happiness than equivalent household income, (6) working status and marital status have a different effect on happiness between men and women.

Keywords JGSS · Age-Period-Cohort analysis · Happiness · Japan · Gender difference

1 Fluctuations in Happiness

Research into happiness is attracting considerable attention in recent years. The movement began with the first publication of the global periodical *Journal of Happiness Studies* in the year 2000; In the field of political science and economics, Bruno Frey and Alois Stutzer's book *Happiness and Economics* was published in 2002; Kwansei Gakuin University launched a research base of "Social Research for the

K. Shishido (✉) · T. Sasaki
JGSS Research Center, Osaka University of Commerce, Higashi-osaka, Japan
e-mail: kuniaki@oak.ocn.ne.jp

T. Sasaki
e-mail: taksasaki@gmail.com

M.-C. Tsai and N. Iwai (eds.), *Quality of Life in Japan*, Quality of Life in Asia 13,
https://doi.org/10.1007/978-981-13-8910-8_2

Enhancement of Human Well-being" in 2003; Zygmunt Bauman wrote *The Art of Life* in 2008; and the manifest released by the Democratic Party of Japan in June 2010 contained the keywords "minimally unhappy society."

This boom in happiness research was sparked by two sets of circumstances. One has to do with the fluctuations in happiness that occur within a wealthy society. As indicated by the results of the World Values Survey, a positive correlation between income and happiness can be observed for countries where the per-capita income is below a certain threshold, but there is no clear association between income and happiness for wealthy societies (Inglehart 2000). "Societies like ours, moved by millions of men and women pursuing happiness, are getting richer, but it is far from clear whether they are getting happier" (Bauman 2009: 9). In Japan, the importance of spiritual happiness surpassed that of material wealth in 1976, following the country's post-WWII period of rapid economic growth (Tomoeda 1996). In affluent societies, an expansion of material wealth no longer explain happiness at both the individual and national levels, Because of this, researchers are taking more interest in what factors determine the happiness of people.

The other is related to the collapse of illusions such as the belief that all Japanese are middle class or that sustained economic growth is possible, combined with a spreading awareness of our increasingly stratified society in which specific traits lead to economic or social poverty and further widen the disparity between its citizens (Yamada 2004; Ishida 2009). This is known as the impasse phase of "affluent societies" (Misawa 1993). Problems created by the gaps in our disconnected society—such as disparities in terms of prospects as well as those associated with a rise in irregular employment among women and youth and increasing numbers of people who never marry—have been identified since the 1990s, and there is a growing perception of the disparities that exist not only in terms of people's economic resources but also in their subjective awareness. That perception is predicated on the realization that a happy society is not one in which people's innate attributes and social opportunities in early life restrict their future life experiences and thus control their happiness.

In this chapter, we use the Age-Period-Cohort (APC) analysis method to investigate the factors that predict happiness in Japanese people, based on cumulative data from the Japanese General Social Surveys (JGSS) conducted ten times from 2000 to 2015. JGSS is a repeated cross-sectional survey, so if we pool the data and employ APC analysis we can examine the period and cohort effects, something which would not be possible using a single-year survey. And if we examine the effects of variables at the individual level while controlling for the period and cohort effects, we can obtain more robust and generalizable findings, not the ones that only hold at one specific point in time.

The following Sect. 2 looks at prior research in the field; Sect. 3 provides an explanation of the analysis method; and Sect. 4 describes the data and variables used in this research. Section 5 contains the analysis of happiness, and finally Sect. 6 presents our conclusions.

2 Prior Research

The predicting factors of happiness are wide-ranging, with reported connections to income, social relationships, personality, and various other individual traits such as age, gender, race, religion, education level, employment status, and health. Here, we will concentrate on the variables used in this research and discuss prior studies related to them.

First is the oft-cited economic factor, which focuses on the relationship between happiness and increases in income. If we look at country-level data in chronological order, we notice that even if income increases, happiness only rises to a certain level and overall does not change that much. But if we take a snapshot of data at the individual level, we see that wealthy people are always happier than those who are not wealthy. This contradiction is known as the Easterlin Paradox (Easterlin 1974). Two ideas that have been put forth to explain this inconsistency are the relative income hypothesis and the adaptation hypothesis[1] (Conceição and Bandura 2008; Firebaugh and Schroeder 2009; Clark et al. 2008; Tsutsui 2010). In addition, the relationship between income and happiness is nonlinear and the resultant effect on happiness decreases as income increases, a phenomenon which is called diminishing marginal utility (Shiraishi and Shiraishi 2010). Di Tella et al. (2001) pointed out that national unemployment and inflation rates have a negative impact on people's happiness.

Next are the predicting factors concerning social relationships. There is a large body of research suggesting that positive personal relationships and a sense of belonging are necessary for maintaining happiness (Diener and Seligman 2004; Frey and Stutzer 2002). Within the family domain, for instance, individuals in both Japan and the United States are happier when they have spouses (Iwai 2002), while men with no spouses exhibit higher levels of psychological distress (Inaba 2002). We also see a study demonstrating that elderly people who live with their unmarried children have low levels of happiness yet those who live with their married children have high levels of happiness (Shishido 2007). In the community domain, people with friends are happier and those who participate in local volunteer activities enjoy higher degrees of satisfaction (Harlow and Cantor 1996; Thoits and Hewitt 2001); areas with higher social capital have also been shown to have happier residents (Putnam 2000).

Gender is one of the individual traits that plays a role, as women tend to be somewhat happier than men (Inglehart 1990; White 1992). There are two points of view regarding age: the U-shaped curve hypothesis and the decline with age hypothesis. Blanchflower and Oswald (2008) examined data for the United States and European countries and reported that those happiness plots traced a U-shaped curve with its lowest points representing people who are in their 40s. Regarding this result, we can recall the observation of Ruth Benedict (1967: 293) that "the life curve of Japan…is a shallow U-shaped curve with a large base, indicating that babies and

[1]The relative income hypothesis states that people's happiness is affected not only by their amount of absolute income but also their amount of income compared to those around them. The adaptation hypothesis states that people's happiness temporarily increases when their income rises but returns to its original level once they become accustomed to their new status.

the elderly are permitted the most freedom and self-indulgence." However, the *White Paper on National Life* noted that the happiness of Japanese people declines as they age (Social Policy Bureau 2008). One problem in identifying happiness transitions that occur as people age is that those shifts can intermingle with ones caused by the cohort effect, and since the measurement scale for happiness (the dependent variable) and the method for controlling variables other than age vary by study, it will take some careful investigation to determine which of these two hypotheses is correct. Iwai (2002) used JGSS-2000 data and found that happiness was higher in older age groups; Iwai and Shishido (2010) then analyzed the happiness changes by gender and age in JGSS cumulative data 2000–2008 and found that the happiness of men in their 60s increased during that eight-year period. In addition to age and gender, other frequently studied individual traits include health, religious faith, and academic background. Health is a strong predicting factor in happiness (Larson 1978). Religious faith can serve as a stress buffer in times of crisis, such as the death of a spouse, unemployment, or poverty, and people who are deeply religious enjoy high levels of happiness (Hamada 2008; Shiraishi and Shiraishi 2010). When we consider academic background, intuitively we think that the more educated a person is the more stable his or her prospects will be, which would positively affect that person's happiness. But higher education can lead to increased ambition and loftier goals, things we expect would have a negative effect on happiness. This may end up offsetting the positive aspects of higher education.

As a last example, let us consider that of Yang (2008), who used the APC method on United States GSS data from 1972–2004 to analyze happiness. Yang's analysis placed period and cohort in the group level and other variables that predict happiness, such as age, in the individual level. Estimates from models that incorporated only age, age squared, period, and birth cohort revealed that although the happiness of Americans slowly increases with age, members of the baby-boomer generation have low happiness. The research also indicated that the differences in happiness due to race decrease with age and that as time goes on the differences due to gender and race are diminishing. The research conducted by Yang is a good reference for how a study can leverage the special properties of repeated cross-sectional surveys to include not only variables attributed to the individual but also period and cohort in the analysis.

Several lines of evidence in Western countries suggest that there are three major factors affecting individual happiness: socio-economic status (education, occupation, income etc.), social relationships (marital status, presence of children, friendship etc.) and health condition. Other findings often show that women are happier than men, and the age effect draws a U-shaped curve with the bottom in middle age (Yang 2008; Blanchflower and Oswald 2004). The effect of some variables on happiness varies as people age. For example, Yang (2008) used American GSS data and found that there is an interaction effect between age and gender. Although women are happier than men in early adulthood, gender difference diminishes as they get older, and men become happier than women in late adulthood. In addition, age and educational background also have a weak interaction effect. The negative effect on happiness due to low educational background weakens in late adulthood and the gap of happiness by educational background decreases. This reduction in the happiness disparity could be attributed to the social security system such as pension and medical care. Another possibility is that the corrosion of social support and social integration (e.g., retirement, separation from children, death of a spouse or friend, etc.) occurs in late adulthood regardless of educational background, gender, or social stratifications. In this paper, we will examine to what extent these findings observed in the West apply to Japan.

3 Analysis Method (Age-Period-Cohort Analysis)

Many indicators used in sociology are time-dependent, from macro-level factors such as the employment rate and the percentage of never married to micro-level factors such as an individual's exercise frequency and subjective health status. These changes that occur over time have attracted great interest in the quest to understand sociological phenomena and have been classified into the three effects: age, period, and cohort. The age effect represents the accumulation of experience and developmental changes that occur as people get older. The period effect represents changes caused by the social, cultural, and economic shifts that take place at particular points in time. And finally, the cohort effect represents the influence of belonging to a specific age group, that is, the unique factors that are imparted on all individuals who went through their development phases at the same time.

The goal of APC analysis is to differentiate these three effects, but their completely linear relationship [age at time of survey = survey year (period) − birth year (cohort)] creates significant problems in terms of methodology. Cumulative individual data for a representative sample has created new possibilities for addressing this challenge. Yang and Land (2006) used the unique qualities of repeated cross-sectional surveys to develop a hierarchical APC model in which individuals belong to their own social context as defined by factors such as birth cohort and survey year. Specifically, the model can control for the effects of age and other variables and estimate the variances and statistical significance of the period and cohort effects by considering the possibility that a group of individuals with the same survey year and cohort share a random error components. This model can be expressed by the following equations:

(1) Level 1 within-cell model:

$$\mathrm{Y}_{ijk} = \beta_{0jk} + \beta_1 \mathrm{AGE}_{ijk} + \beta_2 \mathrm{AGE}_{ijk}^2 + \sum_{m=3}^{m} \beta_3 \mathrm{X}_{3ijk} + \mathrm{e}_{ijk}$$

(2) Level 2 between-cell model:

$$\beta_{0jk} = \gamma_0 + \mathrm{u}_{0j} + \mathrm{v}_{0k}$$

(3) Combined model:

$$Y_{ijk} = \gamma_0 + \beta_1 AGE_{ijk} + \beta_2 AGE^2_{ijk} + \sum_{m=3}^{m} \beta_3 X_{3ijk} + u_{0j} + v_{0k} + e_{ijk}$$

For example, when estimating the happiness Y_{ijk} of individual i, β_{0jk} is the mean value for individuals belonging to the jth birth cohort and kth survey year—in other words, it represents the intercepts of each cell. β_1 through β_m are the fixed effects of each independent variable. When incorporating independent variables other than age into the model, β_{0jk} is the mean happiness for individuals whose independent variable values match the mean value (those who belong to the reference category).[2] Also, e_{ijk} is the difference between the happiness of an individual and the mean value for the cell to which it belongs. In Eq. (2), γ_0 is the mean happiness of the entire sample, u_{0j} is the random effect of birth cohort j on each survey year, and v_{0k} is the random effect of survey year k on each birth cohort. By combining these equations, we can estimate the age effect while taking into account the error components specific to each birth cohort and survey year. The APC analysis method was featured in *Sociological Methods and Research, Volume 36, Number 3* (Smith 2008).

One problem with APC analysis is that although it carries the strict requirement that one of the three factors (age, period, or cohort) must be fixed prior to parameter estimation or relies on macro-level data to estimate their effects, there is still no definitive method for doing so (Glenn 2005; Mason and Wolfinger 2002). The hierarchical APC analysis method used in this study is no exception. We used a limited number of periods and cohorts and the frequencies within the cells were unbalanced, which can cause the estimated parameter values to be skewed. Care should therefore be taken when interpreting our results. If the body of repeated cross-sectional survey data (e.g., JGSS) that is publicly available continues to grow along with the number of periods and cohorts, we will be able to incorporate explanatory variables such as the unemployment rates for those periods and the marital rates for each cohort into our models, which will permit more detailed analyses. And finally, we hope that the methodology behind APC analysis continues to improve and that cumulative data from repeated cross-sectional surveys is adopted by researchers working in a variety of fields.

JGSS includes a happiness variable in self-administered questionnaire form A each time. Subjects are asked to answer the question "Are you happy?" with a number ranging from 1 (Happy) to 5 (Unhappy). The skewed distribution of answers—about 90% fell in the 1–3 range—made us reluctant to treat it as an interval scale (Fig. 1). Therefore, the goal of this study was to investigate the proportion of the Japanese who feels moderately happy, in other words, the rate of individuals who do not feel that they are moderately happy.[3] To that end, we placed those who responded with 1 or 2 in a "happy group" and those who responded with 3, 4, or 5 in an "unhappy

---

[2]The interpretation of intercept values varies depending on the method used to center the independent variables incorporated into a model. In this study, we centered continuous variables using the average values for the entire sample.

[3]Kosaka (2008) suggested that we should focus on reducing unhappiness rather than increasing happiness.

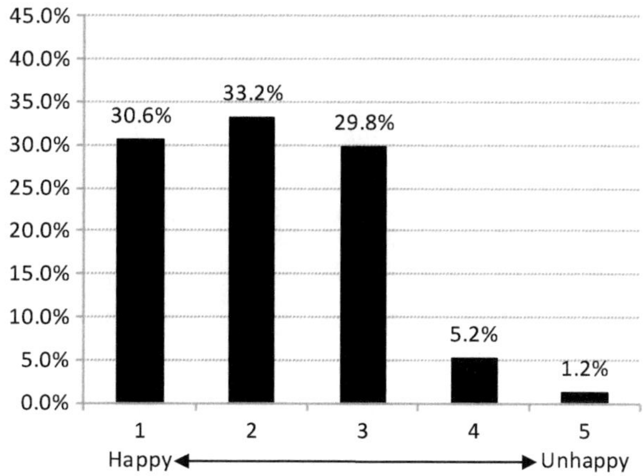

Fig. 1 Happiness distribution (n = 22,266)

group," then converted the happy group to a dummy variable with the value of 1.[4] As a result, we performed a mixed-effect logistic regression for our analysis.

4 Data and Variables

4.1 Data

This study used JGSS cumulative data pooled from ten surveys conducted between 2000 and 2015 (Table 1). JGSS is targeted at men and women between the ages of 20 and 89 who are residents of Japan. Respondents are both interviewed and given a questionnaire to complete on their own. For the surveys conducted in 2003 and from 2006–2012, the sample was randomly divided in half and two types of self-administered questionnaires were used, but the variables used in this study were asked in the same way.

Repeated cross-sectional surveys such as JGSS differ from panel studies in that the sample is reselected each time, which means that they cannot shed light on changes in specific individuals. However, as noted by Micklewright (1994) and Yasuda (2006), the pooled cross-section data compiled from multiple repeated cross-sectional surveys has its advantages: (1) it increases the sample size, (2) it allows us to understand the period effect, (3) it can be used as pseudo-panel data, and (4) it can explain societal changes. Because JGSS is conducted more often than most repeated cross-sectional surveys, it is well-suited for studying the period effect. The number of times a panel study can be conducted is limited due to the need to track down the same individ-

[4]Treating 1–5 as continuous variables and conducting the analysis produces roughly the same results as the ones obtained in this study.

Table 1 JGSS survey details

Conducted date	Survey name	Survey point	No. of valid responses (response rate)
October, 2000	JGSS-2000	300 points	2,893 (64.9%)
October, 2001	JGSS-2001	300 points	2,790 (63.1%)
October, 2002	JGSS-2002	341 points	2,953 (62.3%)
October, 2003	JGSS-2003	489 points	Form A 1,957 (55.0%), Form B 1,706 (48.0%)
August, 2005	JGSS-2005	307 points	2,023 (50.5%)
October, 2006	JGSS-2006	526 points	Form A 2,124 (59.8%), Form B 2,130 (59.8%)
October, 2008	JGSS-2008	529 points	Form A 2,060 (58.2%), Form B 2,160 (60.6%)
February, 2010	JGSS-2010	600 points	Form A 2,507 (62.2%), Form B 2,496 (62.1%)
February, 2012	JGSS-2012	600 points	Form A 2,332 (59.1%), Form B 2,335 (58.8%)
February, 2015	JGSS-2015	300 points	2,079 (52.6%)

uals; since repeated cross-sectional surveys reselect the sample each time, they can be conducted more frequently, which in turn allows for a more detailed analysis.

4.2 Independent Variables

For variables at the individual level, we incorporated the following trait factors: age, age squared, gender, father's occupation when respondent was 15 years old, academic background, employment status, religiosity, and subjective health status. In terms of economic factors, we incorporated both equivalent household income and relative household income (as subjectively compared to the general populace)[5] and compared the influence of each in order to test the relative income hypothesis. If the relative income hypothesis is correct, then we would expect relative household income to have a greater effect than equivalent household income. For social relationship factors, we incorporated marital status, presence of children, and the indicator variable of living alone to represent the family domain. We also incorporated frequency of dining with friends and group participation as social relationship factors that fall outside of the family domain.

The effects that employment status, income, marital status, and social relationships have on happiness are expected to differ by gender (Inaba 2002; Iwai 2002, 2007; Lee

[5]Respondents were asked to answer to the question "Compared with Japanese families in general, what would you say about your family income?" with a number ranging from 1 (Far below average) to 5 (Far above average).

and Ono 2008; Shishido 2007). Accordingly, when incorporating these variables, we divided the sample into men and women before performing our analysis.

The group-level variables are period and cohort. Our periods are the ten points in time that the survey was conducted. The total of seventeen birth cohorts were created by dividing the period between 1910 and 1994 into five-year segments. The definitions and basic statistics for our incorporated variables are provided in Supplementary Table 1, while Supplementary Table 2 indicates the structure of data in this model.

Excluding those who responded to self-administered questionnaire form B, our pooled sample contains 23,718 cases; however, we limited our sample to the 22,266 cases who did not have missing values for the list of variables incorporated into our analysis.

5 Analysis

5.1 Effects of Age, Period, and Cohort on Happiness

Table 2 contains the analysis results for the predicting factors of happiness. The fixed-effect values in the table are odds ratios. Since Models 1 through 6 account for the period and cohort effects, the results for individual-level variables are applicable to individuals in any periods and cohorts. Model 1 incorporates age and age squared at the individual level, with period and cohort variance components introduced as social context. Age squared has a positive and statistically significant relationship with happiness, and the variance components for period and cohort are also significant. As a result, the relationship between age and happiness traces a U-shaped curve; we can say then that happiness varies with period and cohort.

Figure 2 depicts the happiness plots created by calculating the age, period, and cohort effects from the logistic regression coefficients of Model 1. The age effect (a) exhibits a U-shaped curve with a base around 50 years old, which is roughly consistent with the analysis results of Blanchflower and Oswald (2008). Figure 3 shows the age effect for Model 2, which introduces an interaction term between age and gender. Clearly, the age effect differs by gender, with men being less happy than women; the base of the curve is in the early 40s for men but in the early 60s for women. From the period effect illustrated in Fig. 2b, we can see a significant dip in happiness for the year 2003 ($p < 0.01$), then an increase in happiness in 2008 ($p < 0.05$). These changes resemble what would happen if we reversed the transitions in unemployment rate described in the *Labor Force Survey* conducted by the Ministry of Internal Affairs and Communications (Fig. 4), and they match the conclusion of Di Tella et al. (2001) that the national unemployment rate influences people's happiness. The cohort effect in Fig. 2c indicates significantly ($p < 0.1$) low happiness among people who were born from 1935–1939 and thus were children in the years immediately following the Second World War. Growing up in the chaos of the post-war period meant many of

Supplementary Table 1 Variable definitions and basic statistics (n = 22,266)

	Definition of variable	Min.	Max.	Mean	S.D.
Dependent variable happiness	Respondent who answer 1 or 2 in the response category from "Happy" (1) to "Unhappy" (5) is 1. Otherwise is 0	0.00	1.00	0.64	0.48
Level 1 Variable					
Age	Centered to the average, and divided 10	−3.22	3.68	0.00	1.67
Female	Female is 1. Male is 0	0.00	1.00	0.54	0.50
Father's occupation at age 15 (ref. = blue-collar worker)					
Professional and management	Professional or management worker in the SSM occupation 8 classification is 1. Otherwise is 0	0.00	1.00	0.10	0.30
Clerical and selling	Clerical or selling worker in the SSM occupation 8 classification is 1. Otherwise is 0	0.00	1.00	0.22	0.42
Agriculture, forestry and fisheries	Agriculture, forestry and fisheries worker in the SSM occupation 8 classification is 1. Otherwise is 0	0.00	1.00	0.22	0.41
Unemployed and absence	"Father was not working" or "I didn't have father at that time" is 1. Otherwise is 0	0.00	1.00	0.09	0.29
Don't know and no answer	"Don't Know" or "No Answer" is 1. Otherwise is 0	0.00	1.00	0.08	0.27
Educational background (ref. = high school)					
Junior high school and under	Junior high school and under is 1. Otherwise is 0	0.00	1.00	0.20	0.40
University and over	University and over is 1. Otherwise is 0	0.00	1.00	0.33	0.47
Employment status (ref. = regular worker)					
Irregular worker	Daily worker, part-time worker, doing piece work at home, and dispatched worker is 1. Otherwise is 0	0.00	1.00	0.16	0.37
Self-employed and family worker	Self-employed and family worker is 1. Otherwise is 0	0.00	1.00	0.11	0.32
Out of work	Respondent who is out of work and not seeking a job is 1. Otherwise is 0	0.00	1.00	0.34	0.47
Unemployed	Respondent who is out of work and seeking a job is 1. Otherwise is 0	0.00	1.00	0.04	0.20
Religious devotion	"No religion"(0)~"Very devoted"(3). Centered to the average	0.00	3.00	0.49	0.77

(continued)

Supplementary Table 1 (continued)

	Definition of variable	Min.	Max.	Mean	S.D.
Health status	"Bad"(1)~"Good"(5). Centered to the average	1.00	5.00	3.54	1.14
Equivalent household income (ref. = middle quartile)					
Lower quartile (lower 25%)	Lower 25% of annual household income divided by square root of the number of family member is 1. Otherwise 0	0.00	1.00	0.17	0.38
Upper quartile (upper 25%)	Upper 25% of annual household income divided by square root of the number of family member is 1. Otherwise 0	0.00	1.00	0.18	0.38
Refused	Respondent who answer "Don't want to state the income" is 1. Otherwise is 0	0.00	1.00	0.10	0.31
Don't know and no answer	Respondent who answer "Don't know" or Respondent who don't answer is 1. Otherwise is 0	0.00	1.00	0.20	0.40
Subjective household income compared with the others	Response categories are from "Far below average"(1) to "Far above average"(5). Centered to the average	1.00	5.00	2.61	0.86
Marital status (ref. = married)					
Divorced and widowed	Respondent who is divorced or widowed is 1. Otherwise is 0	0.00	1.00	0.12	0.32
Unmarried	Respondent who is unmarried is 1. Otherwise 0	0.00	1.00	0.15	0.36
Having child	Respondent who have child is 1. Otherwise is 0	0.00	1.00	0.78	0.41
Living alone	Respondent who live alone is 1. Otherwise is 0	0.00	1.00	0.08	0.28
Frequency of dine/meet with friends	Frequency of dine/meet with friends per week	0.00	7.00	0.51	1.03
Social group participation	Respondent who belong political associations, trade associations, social service groups, citizen's movement groups, religious groups, sports groups or hobby groups is 1. Otherwise is 0	0.00	1.00	0.40	0.49
Level 2 Variable		Min.	Max.	n	
Period	Survey time point	2000	2015	10	
Cohort	Birth cohort by five-year unit	1910	1994	17	

Supplementary Table 2 Happiness percentages by cohort and survey year (Inside of a Parenthesis is the Number of Respondent)

	2000	2001	2002	2003	2005	2006	2008	2010	2012	2015	Total
1910~1914 birth cohort	88% (24)	56% (18)	60% (5)	75% (4)							73% (51)
1915~1919 birth cohort	63% (46)	69% (64)	68% (44)	73% (26)	50% (16)	75% (8)	100% (3)				67% (207)
1920~1924 birth cohort	70% (110)	68% (91)	68% (97)	58% (62)	60% (45)	78% (37)	73% (33)	79% (28)	50% (16)		68% (519)
1925~1929 birth cohort	65% (186)	66% (177)	62% (160)	57% (128)	64% (96)	66% (103)	74% (66)	74% (61)	70% (56)	83% (29)	66% (1062)
1930~1934 birth cohort	66% (236)	57% (220)	66% (245)	54% (167)	65% (146)	69% (134)	67% (126)	67% (109)	63% (115)	72% (85)	64% (1583)
1935~1939 birth cohort	52% (256)	50% (234)	55% (263)	59% (190)	61% (187)	57% (177)	66% (164)	66% (172)	66% (161)	64% (131)	59% (1935)
1940~1944 birth cohort	55% (286)	59% (269)	59% (295)	55% (195)	56% (213)	62% (213)	67% (172)	66% (245)	69% (216)	67% (157)	61% (2261)
1945~1949 birth cohort	62% (309)	62% (285)	58% (305)	53% (177)	59% (197)	69% (219)	71% (215)	68% (262)	68% (223)	67% (201)	64% (2393)
1950~1954 birth cohort	60% (268)	59% (249)	62% (270)	59% (164)	66% (179)	62% (191)	61% (217)	65% (234)	61% (242)	66% (185)	62% (2199)

(continued)

Supplementary Table 2 (continued)

	2000	2001	2002	2003	2005	2006	2008	2010	2012	2015	Total
1955~1959 birth cohort	65% (217)	61% (205)	60% (242)	56% (160)	64% (152)	57% (165)	66% (190)	54% (190)	69% (189)	68% (174)	62% (1884)
1960~1964 birth cohort	67% (204)	67% (197)	61% (205)	68% (145)	75% (133)	63% (144)	61% (168)	61% (237)	61% (186)	56% (174)	64% (1793)
1965~1969 birth cohort	61% (180)	65% (182)	66% (208)	68% (124)	68% (151)	69% (167)	66% (157)	66% (206)	67% (166)	70% (159)	66% (1700)
1970~1974 birth cohort	61% (208)	71% (170)	63% (199)	60% (128)	66% (148)	68% (184)	73% (176)	70% (217)	74% (216)	67% (168)	67% (1814)
1975~1979 birth cohort	59% (170)	62% (159)	64% (135)	64% (99)	68% (106)	71% (146)	69% (137)	77% (163)	71% (167)	73% (161)	68% (1443)
1980~1984 birth cohort	55% (11)	55% (42)	60% (86)	59% (73)	65% (108)	64% (80)	64% (80)	63% (120)	72% (114)	69% (141)	64% (855)
1985~1989 birth cohort					75% (8)	79% (34)	60% (52)	63% (111)	65% (103)	67% (131)	66% (439)
1990~1994 birth cohort									71% (41)	66% (87)	67% (128)
Total	61% (2711)	61% (2562)	61% (2759)	59% (1842)	64% (1885)	65% (2002)	67% (1956)	66% (2355)	67% (2211)	67% (1983)	64% (22,266)

Table 2 Determining factors of happiness (logistic cross-classified random effect APC models)

Fixed effects	Model 1	Model 2	Model 3	Model 4	Model 5		Model 6	
	All sample (n = 22,266)	All sample (n = 22,266)	All sample (n = 22,266)	All sample (n = 22,266)	Male (n = 10,298)	Female (n = 11,968)	Male (n = 10,298)	Female (n = 11,968)
Intercept	1.69***	1.52***	1.45***	1.39***	1.74***	1.76***	1.87***	2.05
Age	1.00	1.06**	1.07**	1.11***	1.15***	0.99	0.97	0.96*
Age^2	1.02**	1.02**	1.02**	1.02**	1.03**	1.05***	1.09***	1.09***
Female		1.22***	1.22***	1.24***				
Age × Female		0.90***	0.91***	0.93***				
Father's occupation at age 15 (ref. = blue-collar worker)								
Professional and management			1.45***	1.25***	1.10	1.21*	1.09	1.24*
Clerical and selling			1.20***	1.10*	1.09	1.03	1.11	1.03
Agriculture, forestry and fisheries			0.92	0.98	1.14*	0.84**	1.11	0.85**
Unemployed and absence			0.97	0.99	1.23*	0.92	1.29*	0.93
Don't know and no answer			0.85**	0.86**	1.05	0.77**	1.04	0.77**
Age × father's occupation: professional and management			0.99					

(continued)

Table 2 (continued)

Fixed effects	Model 1	Model 2	Model 3	Model 4	Model 5		Model 6	
	All sample (n = 22,266)	All sample (n = 22,266)	All sample (n = 22,266)	All sample (n = 22,266)	Male (n = 10,298)	Female (n = 11,968)	Male (n = 10,298)	Female (n = 11,968)
Age × father's occupation: clerical and selling			1.02					
Age × father's occupation: agriculture, forestry and fisheries			1.02					
Age × father's occupation: unemployed and absence			0.97					
Age × father's occupation: don't know and no answer			0.97					
Educational background (ref. = high school)								
Junior high school and under				0.71***	0.86*	0.99	0.92	1.00
University and over				1.37***	1.16**	1.15**	1.19**	1.18**

(continued)

Table 2 (continued)

Fixed effects	Model 1	Model 2	Model 3	Model 4	Model 5		Model 6	
	All sample (n = 22,266)	All sample (n = 22,266)	All sample (n = 22,266)	All sample (n = 22,266)	Male (n = 10,298)	Female (n = 11,968)	Male (n = 10,298)	Female (n = 11,968)
Age × educational background: junior high school and under				1.07*				
Age × educational background: university and over				0.98				
Employment status (ref. = regular worker)								
Irregular worker					0.67***	0.98	0.87	0.87*
Self-employed and family worker					0.95	1.07	0.98	0.92
Out of work					0.82*	1.44***	0.99	1.22**
Unemployed					0.54***	0.77*	0.74*	0.68**
Religious devotion					1.17***	1.18***	1.14**	1.15***
Health status					1.68***	1.73***	1.73***	1.71***
Equivalent household income (ref. = middle quartile)								

(continued)

Table 2 (continued)

Fixed effects	Model 1	Model 2	Model 3	Model 4	Model 5		Model 6	
	All sample (n = 22,266)	All sample (n = 22,266)	All sample (n = 22,266)	All sample (n = 22,266)	Male (n = 10,298)	Female (n = 11,968)	Male (n = 10,298)	Female (n = 11,968)
Lower quartile (lower 25%)					0.84*	0.79***	0.91	0.88
Upper quartile (upper 25%)					0.95	1.22**	1.08	1.21**
Refused					0.74***	0.90	0.77**	0.92
Don't know and no answer					0.76***	0.82**	0.92	0.89
Subjective household income compared with the others					1.61***	1.56***	1.52***	1.50***
Marital status (ref. = married)								
Divorced and widowed							0.32***	0.87*
Unmarried							0.19***	0.40***
Having child							0.90	0.87
Living alone							0.81*	0.78**
Frequency of dine/meet with friends							1.09**	1.14***

(continued)

Table 2 (continued)

Fixed effects	Model 1	Model 2	Model 3	Model 4	Model 5		Model 6	
	All sample (n = 22,266)	All sample (n = 22,266)	All sample (n = 22,266)	All sample (n = 22,266)	Male (n = 10,298)	Female (n = 11,968)	Male (n = 10,298)	Female (n = 11,968)
Social group participation							1.22***	1.26***
Variance components								
Period effect	0.012***	0.012***	0.011***	0.007***	0.000	0.007**	0.001	0.007**
Cohort effect	0.007***	0.007***	0.007***	0.005**	0.003	0.001	0.004	0.000

*$p < 0.05$, **$p < 0.01$, ***$p < 0.001$

Note Coefficients of fixed effect in this table are odds ratio. Fixed effect logit coefficients, random effect coefficients, standard errors are omitted

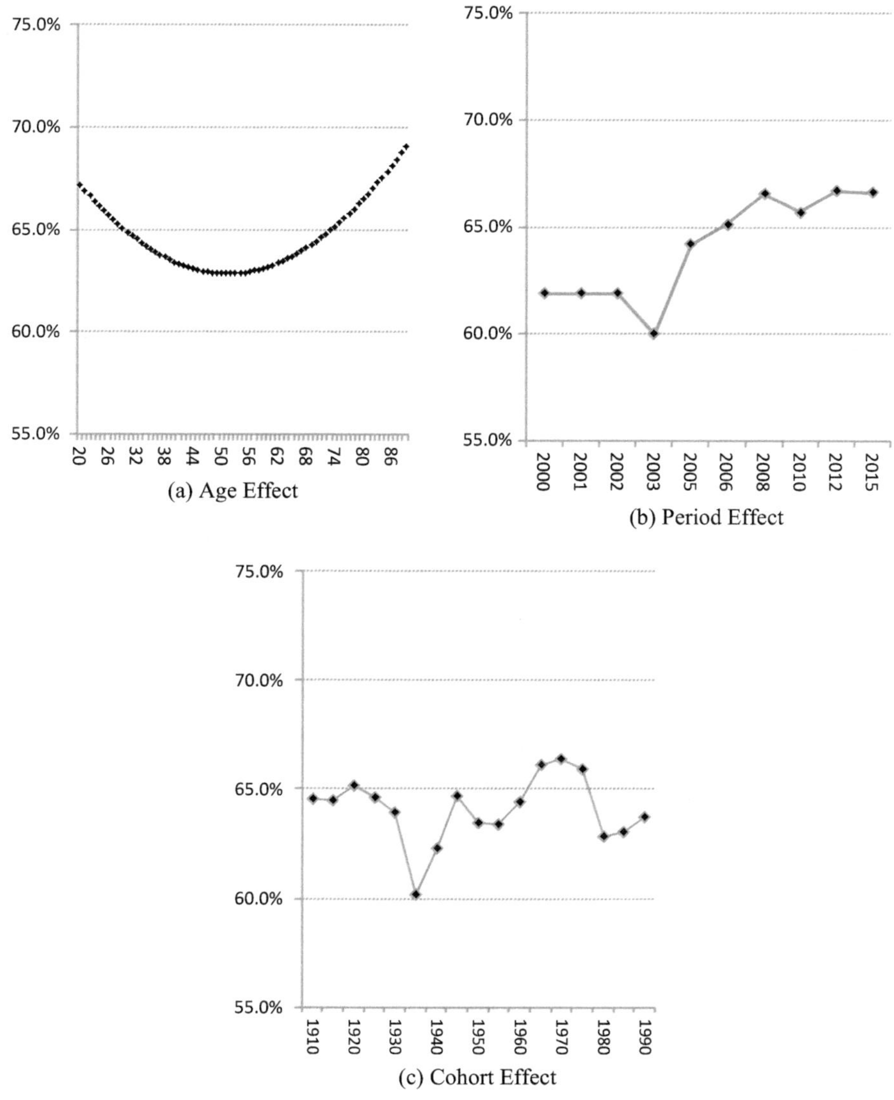

Fig. 2 Estimated effects of age, period, and cohort (from model 1)

these individuals did not have fathers or adequate educational opportunities, factors which presumably had a negative impact on their future lives.[6] Happiness also tended to be lower for people who were born after 1980 because they graduated high school or college after the collapse of the bubble economy and had difficulty finding jobs, but

[6]JGSS-2006 asked about their experiences with giving up on higher education due to financial reasons, and as expected, those born between 1935 and 1939 had the highest rates of abandoning higher education.

Fig. 3 Gender differences in age effect

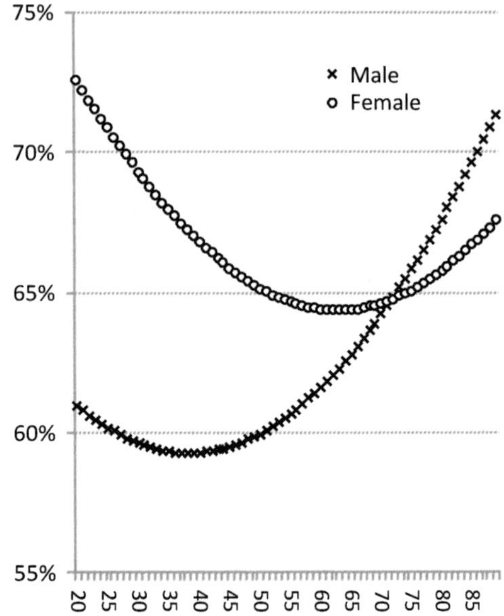

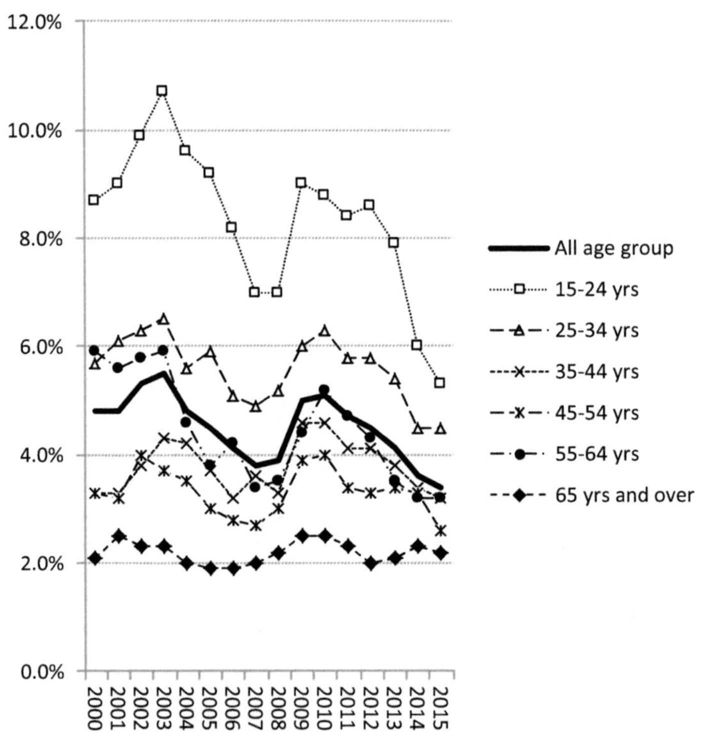

Fig. 4 Changes in unemployment rate

this trend was not statistically significant. It is important to note that Kobayashi (2016) conducted an APC analysis on level of life satisfaction using data from the NHK's *Japanese Value Orientations Survey* (1973–2008) and found that life satisfaction was high among individuals who belonged to the birth cohort of 1979–1993. More research is needed on the happiness and life satisfaction of people born after 1980.

5.2 Happiness and Social Inequality

Models 3 and 4 verify the effects of academic background as well as father's occupation when the individual was 15 years old, two indicators which are connected to the social class of a person's early life. For Model 3, we added father's occupation at 15 and an interaction term between age and father's occupation at 15. Those who had fathers with white-collar jobs at 15 had higher happiness than those with blue-collar fathers, and no significant interaction effect was observed (Fig. 5). Model 4 was created by adding academic background and an interaction term between age and academic background to Model 3 (and omitting the age and father's occupation at 15 interaction term). Because the prestige of an individual's father positively correlates with academic background, incorporating academic background into the model reduced the effects of father's occupation. Plotting the effect of academic background,

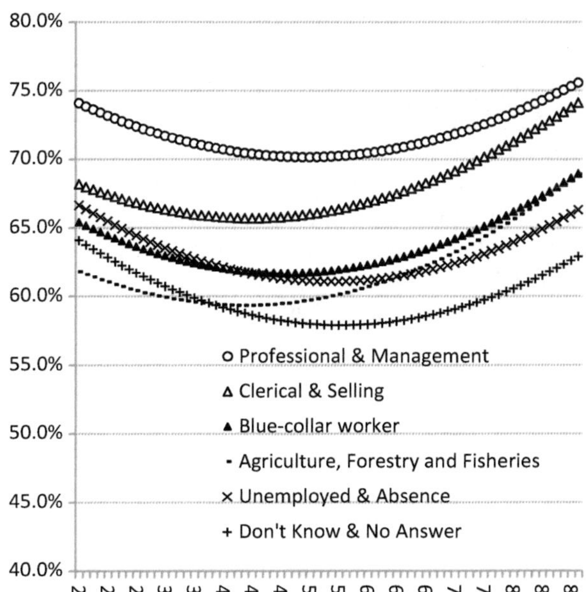

Fig. 5 Happiness and father's occupation

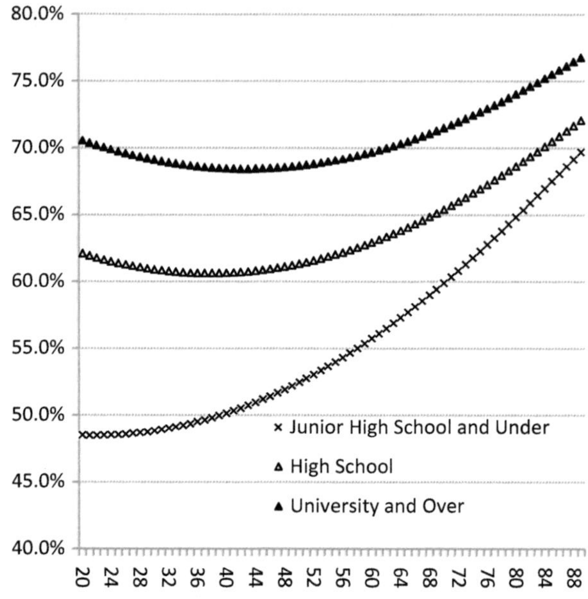

Fig. 6 Happiness and academic background

we see that respondents who have high educational background indicate high level of happiness from their early life (Fig. 6).

The social stratification of an individual in early life, as determined by factors such as father's occupation and academic background, is related to future factors such employment status, income, and marital status. As we can see from the results for Models 5 and 6, introducing factors such as employment status, income, and marital status lessen the effects of these early-life social stratification variables. Looking at how the pattern of happiness changed from Models 3 through 6, as we introduced new variables in the chronological order that life events generally follow, we can see the path [social class at birth → academic background → employment status/income/marital status] and cannot deny the unequal mechanism that surrounds happiness.

5.3 Relative Income Hypothesis and Gender Differences

Model 5 adds employment status, religiosity, and health status to Model 4 and verifies the effects of absolute income (equivalent household income) and relative income (household income as subjectively compared to general society).

The effect that absolute income has on happiness is not as apparent as that of relative income (Fig. 7). There is no difference in happiness between men whose absolute household income is "medium" and those with a "high" income, demon-

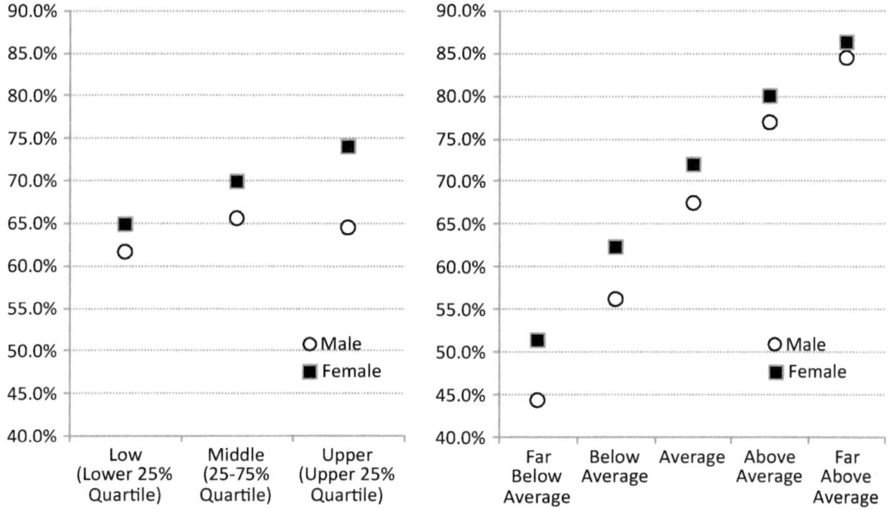

(a) Equivalent Household Income (b) Subjective Relative Income

Fig. 7 Absolute income and relative income (estimated from Model 5)

strating the law of diminishing marginal utility. This is because the long working hours and job-related stress required to earn a high income prevent happiness from increasing. On the other hand, happiness clearly increases with each step on the scale of subjective relative income, meaning that an individual's awareness of his or her own income in relation to the household incomes of others has a profound effect on happiness.

The effect that employment status has on happiness is different for men and women. The happiness of men is lower when they are unemployed, irregularly employed, or lose their jobs. But the happiness of women does not fall as much when they are irregularly employed, and happiness was highest for women who are unemployed because that group consists mostly of housewives. These predicting factors of happiness are rooted in the gender role norm. Also, the variance of period and cohort effects loses statistical significance for men, and cohort effect loses statistical significance for women. We believe this is because period and cohort factors had an effect on individual employment and financial status of men, while only the period effect remained in the case of women. This might indicate that the jobs and income of the many women who return to the workforce after getting married or having a baby are heavily influenced by the state of the economy in those particular periods.

Model 6 introduces factors concerning social relationships into Model 5. For both men and women, the status of social relationships within the family and community domains had a mostly positive correlation with happiness (as social relationships became more abundant, happiness also became higher). But the effects of marital status on happiness varied greatly by gender. Unmarried men or those who experienced divorce or the death of a spouse exhibited large drops in happiness compared to women under the same conditions. Inaba (2002) stated that "men are the ones who gain the most from the psychological benefits of the marital system…and what makes that possible is a societal structure built on gender asymmetry in which women provide the care." This conclusion was supported by our research. When we incorporated marital status into the model, the effect of employment status on men's happiness decreased. This is because the group of irregularly employed or unemployed men is overwhelmingly comprised of unmarried individuals (Table 3). There is also asymmetry between the genders in terms of how easy it is to get married; when men have low income or levels of education, they "won't be chosen" by women (Yamada 2004: 137). So, in addition to the instability created by irregular employment or lack of employment, the unhappiness of men is further increased when they are not selected as a partner under those conditions. If we control for marital status, the linear effect of age disappears for men and instead becomes a U-shaped curve approximately the same as that for women. The sag in the first half of the U-shaped curve for men possibly represents the unhappiness of unmarried youth who have not yet found stable employment.

6 Conclusion

In this study, we used the APC analysis method on JGSS cumulative data 2000–2015 to investigate the effects that age, period, and cohort have on happiness and analyze the effects of individual-level variables while controlling for period and cohort. Our results can be summarized by the following six points: (1) the age effect traces a U-shaped curve; (2) happiness hit its lowest level in 2003, which also had the highest unemployment rate in this fifteen-year period; (3) the 1935–1939 birth cohort had low happiness; (4) an individual's social class at birth and social opportunities in early life affect how happiness changes as he or she ages; (5) relative income has a stronger correlation with happiness than absolute income does; and (6) the effects that employment and marital status have on happiness differ for men and women, with men experiencing substantial unhappiness when they lack a spouse.

There are three noteworthy things in these results. First of all, the distribution of subjective happiness over people's lives can be either directly or indirectly affected by their social classes at birth and in early life. Since the occupation of an individual's father affects his or her academic background, which then goes on to affect factors such as employment status, income, and marital status, we can't rule out the possibility that those early indicators influence happiness.

Second, happiness is affected much more strongly by people's awareness of how their incomes are ranked relative to those of the people around them (the reference group) than it is by the objective size of their income. This result supports the relative income hypothesis of Firebaugh and Schroeder (2009). Also, the fact that equivalent household income and the happiness of men have a nonlinear relationship indicates that happiness from income is subject to diminishing marginal utility, which partially supports the adaptation hypothesis.

Table 3 Cross table of employment status and marital status (ages 20–64) (Row %)

Male		Married (%)	Divorced and widowed (%)	Unmarried (%)	n	Female		Married (%)	Divorced and widowed (%)	Unmarried (%)	n
20–34 yrs	Regular worker	49.2	1.1	49.6	1,265	20–34 yrs	Regular worker	27.8	2.6	69.6	832
	Irregular worker	7.1	0.0	92.9	240		Irregular worker	44.1	5.6	50.4	540
	Self-employed and family worker	49.6	2.5	47.9	119		Self-employed and family worker	62.0	5.6	32.4	71
	Out of work	3.8	0.8	95.5	133		Out of work	78.3	0.7	21.1	589
	Unemploye	9.6	1.1	89.4	94		Unemploye	50.9	1.8	47.2	163
35–49 yrs	Regular worker	82.4	3.7	13.8	2,022	35–49 yrs	Regular worker	67.2	14.8	18.0	818
	Irregular worker	37.0	8.2	54.8	73		Irregular worker	87.6	7.1	5.3	1,106
	Self-employed and family worker	76.1	5.3	18.7	284		Self-employed and family worker	90.0	4.6	5.4	241
	Out of Work	28.6	3.6	67.9	28		Out of Work	94.9	1.8	3.3	719
	Unemploye	33.9	3.6	62.5	56		Unemploye	89.6	3.3	7.1	182

(continued)

Table 3 (continued)

50–64 yrs	Regular worker	91.1	4.3	4.6	1,852	50–64 yrs	Regular worker	74.5	19.7	5.7	699
	Irregular worker	84.4	8.2	7.4	243		Irregular worker	83.9	13.0	3.1	936
	Self-employed and family worker	88.8	6.8	4.5	605		Self-employed and family worker	87.9	9.6	2.6	470
	Out of work	78.4	9.0	12.6	356		Out of work	85.9	10.7	3.4	1,276
	Unemploye	76.4	9.3	14.3	161		Unemploye	79.6	16.0	4.3	162

Third, there is the issue of unhappiness among young men. The model that incorporates age and age squared contains a dip in the first half of men's U-shaped curve, which may be caused by unhappiness related to unstable employment and their unmarried status. If we take the results of the Labor Force Survey conducted by Ministry of Internal Affairs and Communications for the years 1970–2015, set the unemployment rate for all men to 1, then view the unemployment rate by age group, we can see that while men between the ages of 55 and 64 maintained their employment around the time the bubble collapsed, employment for men aged 15–24 and 25–34 was suppressed (Fig. 8).

How these findings in Japan can be compared to those in Western countries? Major factors that have been found to influence individual happiness, namely, socioeconomic status, social relationships, and health conditions were also found to predict happiness in Japan. A salient difference between the present study and results from Yang (2008) targeting the United States is the interaction effect between age and gender (Fig. 3). More specifically, gender difference in happiness among Japanese is larger than that of the United States, and young men's happiness is significantly lower than women's happiness in Japan. In Japan, gender roles are still rigid both in attitudes and behavior, and a gender gap is quite dramatic compared to other advanced societies. Although Japanese young men are strongly expected to enact breadwinning roles, many men are unable to fulfill such roles due to unstable labor

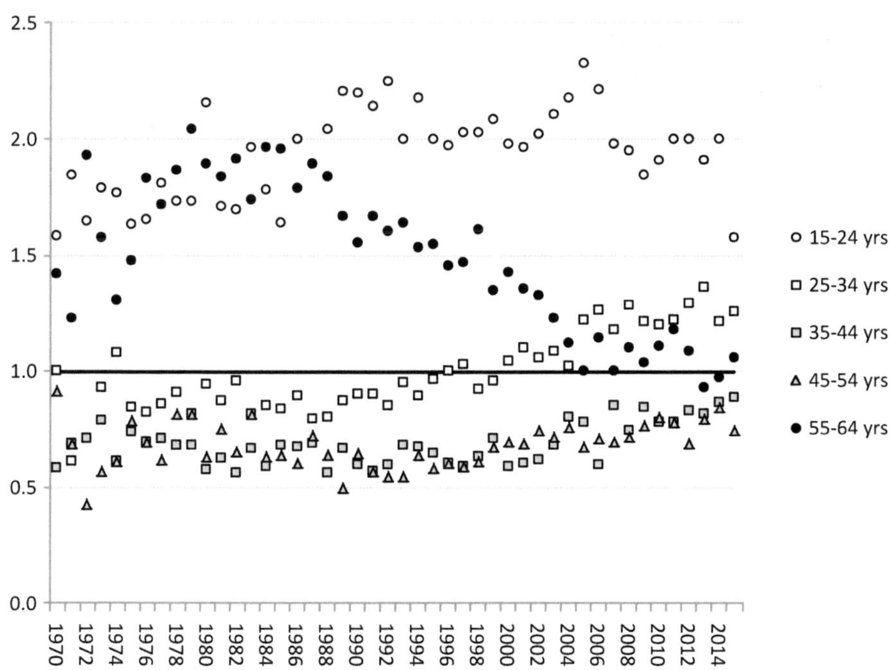

Fig. 8 Unemployment by age group (total male unemployment rate = 1) (results of the *Labor Force Survey* conducted each April by the Ministry of Internal Affairs and Communications)

market. Moreover, young regular employees are often forced to work long hours with low wage under Japanese employment practices including seniority system. These working conditions may have a negative impact on happiness among Japanese young men.

Another difference between the present study and Yang (2008) is the interaction effect between age and educational background (Fig. 6). The disparity in happiness based on educational backgrounds is large in Japan in early adulthood. A possible explanation for this may be the lack of social support for families with small children. Since 1970, as population aging gained notice, social security for the elderly has expanded. In recent years, child poverty (especially among female single-parent families) and intergenerational linkage of poverty are serious social problems in Japan. Because the proportion of students attending private universities is extremely high, and the public scholarship system for university students is also poor, the economic conditions of families are strongly related to the advancement rate for higher education. Thus, the present results suggest that the relationship among economic conditions of parents, children's educational background and children's occupation is strong in Japan.

There is a general sense that the predicting factors of happiness in "affluent societies" shift from tangible to intangible, or in other words, from economic factors to those revolving around social relationships. But our analysis has implied that relative income level (as compared to that of others) plays a tremendous role in shaping people's happiness, and as we saw with the relationship between employment status and marital status, economic factors and social relationship factors are intertwined. It is as though the possibility of happiness being completely divorced from possessions (economic factors) is nothing more than an illusion in "affluent societies".

Acknowledgements The Japanese General Social Surveys (JGSS) are designed and carried out by the JGSS Research Center at Osaka University of Commerce (Joint Usage/Research Center for Japanese General Social Surveys accredited by Minister of Education, Culture, Sports, Science and Technology), in collaboration with the Institute of Social Science at the University of Tokyo. JGSS-2015 is financially assisted by JSPS KAKENHI Grant Number 26245060. This study received funding from MEXT's Grants-in-Aid for Scientific Research (Young Researcher (B) 21730434).

References

Bauman, Z. (2009). *Kōfukuron—"Ikizurai" jidai no shakaigaku* [The art of life] (R. Takahashi & F. Hirakiuchi, Trans.) Japan: Shakuhinsha.

Benedict, R. (1967). *Kiku to katana—Nihon bunka no kata* [The chrysanthemum and the sword: Patterns of Japanese culture] (M. Hasegawa, Trans.). Japan: Shakai Shisousha.

Blanchflower, D. G., & Oswald, A. J. (2004). Well-being over time in Britain and the USA. *Journal of Public Economics, 88,* 1359–1386.

Blanchflower, D. G., & Oswald, A. J. (2008). Is well-being U-shaped over the life cycle? *Social Science and Medicine, 66*(8), 1733–1749.

Clark, A. E., Frijters, P., & Shields, M. A. (2008). Relative income, happiness, and utility: An explanation for the Easterlin paradox and other puzzles. *Journal of Economic Literature, 46*(1), 95–144.
Conceição, P., & Bandura, R. (2008). Measuring subjective wellbeing: A summary review of the literature. *UNDP Working Paper*. Retrieved from www.undp.org/developmentstudies/researchpapers.shtml.
Diener, E., & Seligman, M. E. P. (2004). Beyond money: Toward an economy of well-being. *Psychological Science in the Public Interest, 5*(1), 1–31.
Di Tella, R., MacCulloch, R. J., & Oswald, A. J. (2001). Preferences over inflation and unemployment: Evidence from surveys of happiness. *American Economic Review, 91*(1), 335–341.
Easterlin, R. A. (1974). Does economic growth improve the human lot? In P. A. David & M. W. Reder (Eds.), *Nations and households in economic growth: Essays in honor of Moses Abramovitz* (pp. 89–125). New York: Academic Press.
Firebaugh, G., & Schroeder, M. B. (2009). Does your neighbor's income affect your happiness? *American Journal of Sociology, 115*(3), 805–831.
Frey, B. S., & Stutzer, A. (2002). *Kōfuku no seiji-keizaigaku—Hitobito no shiawase o sokushin suru mono wa nanika* [Happiness and economics: How the economy and institutions affect human well-being] (T. Sawa & F. Sawasaki, Trans.). Japan: Diamond.
Glenn, N. D. (2005). *Cohort analysis*. Thousand Oaks, CA: SAGE.
Hamada, H. (2008). Kōfukukan no genjō [Current state of happiness]. In K. Kosaka (Ed.), *Kōfuku no Shakai Riron* (pp. 78–87). Japan: The Society for the Promotion of the Open University of Japan.
Harlow, R. E., & Cantor, N. (1996). Still participating after all these years: A study of life task participation in later life. *Journal of Personality and Social Psychology, 71*, 1235–1249.
Inaba, A. (2002). Kekkon to disutoresu [Marital status and psychological distress in Japan]. *Shakaigaku Hyōron, 53*(2), 69–83.
Inglehart, R. (1990). *Culture shift in advanced industrial society*. Princeton, NJ: Princeton University Press.
Inglehart, R. (2000). Globalization and postmodern value. *The Washington Quarterly, 23*(1), 215–228.
Ishida, J. (2009). Kasōteki shotokusai bunpai ni yoru fubyōdō to kōfuku sōwa no hendō—2005-nen SSM chōsa dēta o mochiita shimyurēshon bunseki [Change in inequality and aggregation of subjective well-being by redistribution of income: A simulation analysis using the 2005 SSM survey dataset]. *Shakaigaku Hyōron, 59*(4), 752–768.
Iwai, H. (2007). Kōreisha no shakaiteki chii no henka to kōfukukan—Raifu kōsu to kaisō kenkyū no shiten kara [Changes in social status and happiness of the elderly—From the viewpoint of research into life course and class]. *Ikigai Kenkyū, 13*, 47–72.
Iwai, N. (2002). Shiawase desu ka—nihonjin no kōfukukan [Are you happy?—The happiness of Japanese people]. In N. Iwai & H. Satō (Eds.), *Nihonjin no Sugata* (pp. 2–8). Japan: Yuhikaku Publishing.
Kobayashi, D. (2016). Seikatsu manzokukan ni taisuru karei kōka/ cōhōto kōka/ jidai kōka [Effects of aging, period, and cohort on life satisfaction]. In H. Tarōmaru (Ed.), *Kōki Kindai to Kachi Ishiki no Henyō—Nihonjin no Ishiki* (pp. 1973–2008). Tokyo, Japan: University of Tokyo Press.
Kosaka, K. (2008). Fukō no genzan [Reducing unhappiness]. In K. Kosaka (Ed.), *Kōfuku no Shakai Riron*. Tokyo: Foundation for the Promotion of the Open University of Japan.
Larson, R. W. (1978). Thirty years of research on the subjective well-being of older Americans. *Journal of Gerontology, 33*(1), 109–125.
Lee, K. S., & Ono, H. (2008). Specialization and happiness in marriage: A U.S.-Japan comparison. *Social Science Research, 37*, 1216–1264.
Mason, W. M., & Wolfinger, N. H. (2002). Cohort analysis. In N. J. Smelser & P. B. Baltes (Eds.), *International encyclopedia of the social & behavioral sciences* (pp. 2189–2194). St. Louis, MO: Elsevier.

Micklewright, J. (1994). The analysis of pooled cross-section data: Early school leaving. In A. Dale & R. B. Davies (Eds.), *Analyzing social & political change: A casebook of methods* (pp. 78–97). London: SAGE.
Misawa, K. (1993). Gendai nihonjin no ikikata [Modern Japanese way of life]. In K. Misawa (Ed.), *Gendai Seikatsu to Ningen* (pp. 25–47). Tokyo, Japan: Koyo Shobo.
Putnam, R. D. (2000). *Bowling alone: The collapse and revival of American community*. New York, NY: Simon & Schuster.
Shiraishi, K., & Shiraishi, S. (2010). Kōfuku no keizaigaku no genjō to kadai [A survey of happiness research]. In F. Ōtake, S. Shiraishi, & Y. Tsutsui (Eds.), *Nihon no Kōfukudo* (pp. 9–32). Tokyo, Japan: Nippon Hyoronsha.
Shishido, K. (2007). Kōreiki ni okeru kōfukukan kitei yōin no danjosa ni tsuite—JGSS-2000/2001 tōgō dēta ni motozuku kentō [Difference between men's happiness and women's happiness in later life—An analysis based on JGSS integrated data 2000/2001]. *JGSS Kenkyū Ronbunshū, 6,* 45–56.
Shishido, K., & Iwai, N. (2010). JGSS ruiseki dēta 2000–2008 ni miru Nihonjin no ishiki to kōdō no henka [Trends of Japanese values and behavioral patterns based on JGSS cumulative data 2000–2008]. *JGSS Kenkyū Ronbunshū, 10,* 1–22.
Social Policy Bureau. (2008). *Heisei 20 Nenban Kokumin Seikatsu Hakusho [Citizen life white paper for 2008]*. Tokyo, Japan: Cabinet Office.
Smith, H. L. (2008). Advances in age-period-cohort analysis. *Sociological Methods & Research, 36*(3), 287–296.
Thoits, P. A., & Hewitt, L. N. (2001). Volunteer work and well-being. *Journal of Health and Social Behavior, 42,* 115–131.
Tomoeda, T. (1996). Shakaigaku to Jisshōshugi [Sociology and positivism]. *Shakaigaku Hyōron, 47*(1), 33–46.
Tsutsui, Y. (2010). Kōfuku kenkyū no kadai [Challenges in happiness research]. In F. Ōtake, S. Shiraishi, & Y. Tsutsui (Eds.), *Nihon no Kōfukudo* (pp. 263–279). Tokyo, Japan: Nippon Hyoronsha.
White, J. M. (1992). Marital status and well-being in Canada. *Journal of Family Issues, 13,* 390–409.
Yamada, M. (2004). *Kibō kakusa shakai—'Makegumi' no zetsubōkan ga Nihon o hikisaku [A society separated by hope—How the hopelessness of 'losers' disrupts Japan]*. Tokyo, Japan: Chikumashobo.
Yang, Y. (2008). Social inequalities in happiness in the United States 1972 to 2004: An age-period-cohort analysis. *American Sociological Review, 73,* 204–226.
Yang, Y., & Land, K. C. (2006). A mixed models approach to the age-period-cohort analysis of repeated cross-section surveys, with an application to data on trends in verbal test scores. *Sociological Methodology, 36,* 75–97.
Yasuda, T. (2006). Hanpuku ōdan chōsa to shite no JGSS no igi—Chōsa kankaku no mijikasa ga motsu riten [Significance of Japanese GSS as repeated cross-sectional survey—Advantage offered by short survey intervals]. *Shakaigaku Nenshi, 47,* 19–33.

Kuniaki Shishido is a professor of the faculty of public policy, Osaka University of Commerce. He is a member of the JGSS Steering Committee. His areas of specialty are Social Gerontology, Social Survey and Lifestyle. He also participates in cross-cultural survey projects such as the East Asian Social Survey (EASS). His articles have appeared in Asian Journal for Public Opinion Research, Journal of Japanese Family Sociology, International Journal of Japanese Sociology. His recent books include Quality of Life in Late Adulthood: Happiness, Social Network and Civic Activities (Koyoshobo, 2018), The SAGE Handbook of Survey Methodology (co-author, SAGE 2016) and Family Values in East Asia (co-author, Nakanishiya 2011).

Takayuki Sasaki is an Associate Professor in the Faculty of Public Affairs, Osaka University of Commerce, Japan. He has been serving on the steering committee of the JGSS Research Center. The Center is the Joint Usage/Research Center accredited by the Minister of Education, Culture, Sports, Science and Technology of Japan (MEXT), and conducts the Japanese General Social Survey, a biennial large-scale repeated survey with a nationally representative sample. His primary interest lies in Evidence-Based Policy Making (EBPM). His articles have appeared in *Personal Relationships*, *Family Relations*, *Journal of Family Psychology*, and book chapters in *Healthy Aging in Sociocultural Context* (Routledge 2012) and *Parent-Child Relationships of Three Generations: An Empirical Examination using Matching Data* (coeditor, Kazamashobo 2018).

Income, Poverty and Happiness

Happiness in Contemporary Japan: Study of Lifestyle and Values Using the Relative Income Hypothesis

Hiroo Harada and Eiji Sumi

Abstract We conducted a Web Survey on Lifestyle and Values for 11,814 respondents in Japan in February 2015. In this study, we examined the relative income hypothesis, using Japanese nationwide survey data. The main findings of the study were as follows. First, our empirical results provide support for the relative income hypothesis, according to which individual life satisfaction depends not only on one's own income but also on one's income relative to those of others in area where an individual has his or her residence. Second, it was shown that discrete changes in both absolute and relative income had relatively smaller marginal effects on predicted probability of choosing an outcome of life satisfaction than those of discrete changes in individual attributes such as age, marital status, and occupational status, on average. Our contribution to the study of happiness and well-being is that our results support the relative income hypothesis in Japan: individual life satisfaction depends not only on one's own income but also on one's income relative to average income of the municipality where an individual has his or her residence. In addition, we found diminishing marginal utility with relative income. When relative income was sufficiently large, its marginal effect of predicted probability only slightly increased.

Keywords Relative income hypothesis · Life satisfaction · Reference group · Ordered logit · Marginal effect

H. Harada (✉)
Senshu University, Tokyo, Japan
e-mail: h2@isc.senshu-u.ac.jp

E. Sumi
Niigata University, Niigata, Japan
e-mail: esumi@econ.niigata-u.ac.jp

M.-C. Tsai and N. Iwai (eds.), *Quality of Life in Japan*, Quality of Life in Asia 13,
https://doi.org/10.1007/978-981-13-8910-8_3

1 Introduction

1.1 Research Question

Those with higher incomes have greater opportunities to obtain material goods and achieve higher status in society. Thus, higher income yields higher happiness. Economic studies on the determinants of individual life satisfaction (LS) or happiness have focused on subjects' own incomes and occupational status as well as their individual attributes (e.g., Clark and Oswald 1994). However, because a diminishing marginal utility appears with absolute income, the relationship between income and happiness is nonlinear. Higher income does not always bring proportionate higher happiness, because individuals compare themselves to other individuals. In short, it is not the absolute level of income that matters most but rather one's position relative to other individuals. Easterlin (1974) pointed out that people with higher incomes are happier, but raising their income does not increase their happiness, because in comparison to others, their income has not improved (also see Stiglitz et al. 2010).

This relation between relative income and happiness has been the main subject of empirical studies (see Frey and Stutzer 2002). The relative income hypothesis states that individual utility depends both on one's own income and income relative to others (reference group).[1] However, there is still no consensus on the effects of relative income, although one's own income has been found to have a significantly positive effect on utility. Table 1 shows main prior studies that define the reference group as a spatial group (such as those in one's neighborhood, region, municipality, township, county, or state). As shown in Table 1, prior studies that have conducted empirical analysis of the association between utility and relative income have had mixed results. Shields et al. (2009) and Mangyo and Park (2011) found no association in Australia and China, whereas Luttmer (2005) and Helliwell and Huang (2005) found a negative association in the United States and Canada; further, Blanchflower and Oswald (2004) and Clark et al. (2009) found a positive association in the United States and Denmark.

Previous empirical studies of the determinants of happiness in Japan include only absolute income among the independent variables with no consideration for income relative to reference groups. In addition, although previous empirical studies related with relative income hypothesis in Japan examined individual's happiness and the Gini coefficient which indicates the income inequality in one's prefecture or municipality, there is also no consensus on the effect of income inequality in one's place of residence (e.g., Oshio and Kobayashi 2009, 2011; Inaba et al. 2015; Sumi 2016). Thus, in this study, we empirically examined the relative income hypothesis, based on a nationwide survey data of Japanese citizens and tax agency data from the Japanese government.

[1] This hypothesis has been established as the foundation of the consumption function since it was set forth and developed by Duesenberry (1949) and Friedman (1957).

Table 1 Summary of main studies of relative income

Study /country /survey period	Utility proxy /model	Reference group	RI effect
Blanchflower and Oswald (2004) United States 1972–1998	Happiness OLM[e]	-Mean income in state	- Positive
Luttmer (2005) United States 1987–88,1992–94	Happiness FEM[d]	- Mean earnings in PUMA (PUMA = public use microdata area)	- Negative
Clark et al. (2009) Denmark 1994–2001	Satisfaction[a] FEM[d]	- Median income in neighborhood - Median income in municipality - Relative rank in neighborhood	- Positive - Negative but no[g] - Positive
Shields et al. (2009) Australia 2001	LS[b] OPM[f]	- <50% of median income in CD - >200% of median income in CD (CD = census collection district)	- Negative but no[g] - Positive but no[g]
Helliwell and Huang (2005) Canada 2002, 2003	LS[b] OPM[f]	- Mean household income in CT (CT = census tract)	- Negative
Mangyo and Park (2011) China 2004	SRH[c] FEM[d]	- Mean income in township - Mean income in province	- Negative but no[g] - Positive but no[g]
This study Japan 2015	LS[b] OLM[e]	- Mean income in municipality - <50% of mean income - >200% of mean income - >400% of mean income	- Positive - Negative but no[g] - Positive - Positive

Note (1) [a]Satisfaction with economic condition, [b]Life satisfaction, [c]Self–rated health

(2) [d]Fixed effect model, [e]Ordered logit model, [f]Ordered probit model

(3) [g]Notstatistically significant

1.2 Survey Design and Survey Data[2]

We used individual data from the International Comparative Survey on Lifestyle and Values (survey of Japan), conducted in 2015 by the Center for Social Well-being Studies at Senshu University.

The data were derived from an online survey of respondents aged 20–69 years, who were randomly chosen from the pre-registered monitors of a survey agent (Nikkei Research Inc.). Participants were randomly selected using 240 stratified categories constituted in proportion to national distribution with regard to age, gender, city size, and region (see Table 2). Populations were sampled proportionately from 2010 national census findings by gender, age, city scale, and region and were considered to be representative of Japan as a whole. There were 11,814 valid responses.

Table 2 Sample distribution (Unit: %)

Category		Sample	2010 Census
Gender	Male	50.0	49.9
	Female	50.0	50.1
Age	20s	16.3	16.5
	30s	22.0	21.8
	40s	20.0	20.2
	50s	19.5	19.6
	60s	22.1	21.9
Municipality size	Towns and villages	8.3	8.9
	Small cities (population < 200,000)	37.7	37.7
	Medium-sized cities (population 200,000–700,000)	24.9	24.8
	20 major cities (population > 700,000)	29.1	28.7
Region	Hokkaido and Tohoku	11.3	11.3
	Kanto	34.5	34.6
	Chubu	16.8	18.1
	Kinki	18.0	16.4
	Chugoku and Shikoku	8.6	8.6
	Kyushu and Okinawa	10.8	11.0

[2] At Senshu University Center for Social Well-Being Studies (2014–2018), headed by Hiroo Harada, we are conducting an International Comparative Survey on Lifestyle and Values to facilitate a large-scale systematic exploration of the determinants of well-being. Surveys were completed by respondents in Japan (February 2015), South Korea (summer 2015), Vietnam (autumn 2015), the Philippines (autumn 2016), Thailand (winter 2016), and Indonesia and Taiwan (2017). In this study, we analyze the findings from the survey in Japan, completed in February 2015.

The outline of the paper is as follows. In Sect. 2, we describe the relative income. Section 3 explains the model and data. Sections 4 and 5 discusses the results of ordered logit estimation and the marginal effects, and Sect. 6 summarizes our conclusions.

2 Relative Income

In this study, we examined whether individuals are more likely to be satisfied with their lives when they have larger incomes than the average income of their municipality. Therefore, we defined the reference group as inhabitants of the municipality where the individual has his or her residence. The average income of the municipality was taken to be the mean of taxable income in each region of local individual income tax in 2015.[3] For absolute individual income,[4] we adopted the equivalized income calculated by dividing household income[5] by the square root of household size.

We defined the relative income metric for individuals by dividing the equivalized income by municipal average taxable earned income, as follows:

$$\text{Relative income metric} = \text{equivalized income/municipal average income}$$

The scale for relative income theoretically ranges from 0 to infinity, with 1 representing an equivalized income equal to the municipal average income. As shown in Table 3, the numerical values of relative income ranged between 0.093 and 36.240 for all samples. The descriptive statistics for each of the quartiles of relative income (low: the first quartile, lower middle: the second quartile, higher middle: the third quartile and high: the fourth quartile) are also given in Table 3. Because the mean and median values for the low-level group were 1.085 and 1.176 respectively, the

[3]Therefore, taxable income data do not include income earned by those whose income falls below the exemption limits, a category that accounts for almost 40% of income earners in Japan. The proportion of non-taxpayers in a region might differ slightly between regions, but in this analysis the differences are disregarded for the sake of simplicity.

[4]It is inappropriate to adopt individual income as the indicator of own income. This is because even if homemakers don't have own individual income, it doesn't mean that they don't have any income; they live off the income of their husband (i.e., household income). Thus, we adopted the equivalized income, which refers to household income that has been recalculated to take into account differences in household size, as the indicator of own income and the valid measure of individual's current living standards.

[5]In our survey, household income is categorized into 25 income segments, ranging from under 500,000 yen to 20,000,000 yen or more. Median values are used as the reference values for all segments, except for the segment 20,000,000 yen or more, for which 32,600,000 yen is used. According to another data base; the Statistical Survey of Actual Status of Salary in the Private Sector conducted by the National Tax Agency, the average income of income earners who have 20,000,000 yen or more was 32,600,000 yen in 2015.

Table 3 Descriptive statistics for relative income

Relative income	N	Min	Mean	Median	Max
Low	2,605	0.093	1.085	1.176	1.620
Lower middle	2,603	1.621	2.009	2.003	2.412
Upper middle	2,604	2.412	2.901	2.875	3.477
High	2,604	3.477	5.250	4.530	36.240
All sample	10,416	0.093	2.811	2.412	36.240

respondents in our survey could be presumed to be relatively wealthy compared to the average inhabitants of their municipalities.[6]

To clarify the crude relationship between individual LS or happiness and level of relative income,[7] we indicated the distribution of respondents' LS for each relative income level; these data are illustrated in Table 4. As shown in that table, the mode value of LS was found to increase from 5 to 8 as relative income increased from a low

Table 4 Relative income and distribution of life satisfaction (0–10) (Unit: %)

LS	RI				Total
	Low	Lower middle	Upper middle	High	
	1st quartile	2nd quartile	3rd quartile	4th quartile	
0	7.1	2.2	1.8	2.2	3.3
1	3.3	1.8	1.2	1.1	1.9
2	6.8	4.2	3.0	2.8	4.2
3	10.3	7.3	6.8	5.0	7.3
4	8.7	8.3	7.0	5.1	7.3
5	18.6	17.2	16.7	14.4	16.7
6	12.4	12.9	13.3	12.2	12.7
7	13.1	18.8	19.7	19.4	17.7
8	11.1	17.1	18.7	23.3	17.6
9	5.1	5.6	6.9	8.4	6.5
10	3.6	4.7	4.8	6.2	4.8
Total	100.0	100.0	100.0	100.0	100.0
Mode	5	7	7	8	7
Median	5	6	7	7	6

Note N = 10,416

[6]We conducted a preliminary analysis about relative income using our sample data. For more information, refer to Harada (2016a, b).

[7]The correlation between happiness and LS from all samples of the survey results was rather high: the correlation factor was 0.836.

to a high level. The median value of LS also increased from 5 to 7 as relative income increased. Thus, we expected that relative income would have a positive effect on LS.

3 Model and Data

3.1 Model

To analyze the relative income hypothesis, we estimated an ordered logit regression model as follows[8]:

$$LS_i^* = \alpha + \sum_{m=1}^{l} \beta_k X_i^m + \delta EI_i + \pi RI_{ij} + \varphi W_{ij} + u_i \tag{1}$$

here, LS^* was a latent unobserved variable. *LS* was recorded on an 11-point Likert scale (10 = very satisfied, 0 = very unsatisfied) and had 10 thresholds over the latent variables.

$$\mathrm{LS}_i = \begin{cases} 0 \;\; if \; -\infty < LS^* \le k_1 \\ 1 \;\; if \; k_1 < LS^* \le k_2 \\ 2 \;\; if \; k_2 < LS^* \le k_3 \\ \vdots \;\; \vdots \;\; \vdots \\ 10 \;\; if \; k_{10} < LS^* < \infty \end{cases}$$

here, *X* was an individual-level variable, *EI* was equivalized income, *RI* was relative income, *W* was a vector of regional-level variables, *u* was the error term,[9] and α, β, δ, π, φ, and the 10 threshold values *k* were the parameters to be estimated. Moreover, *i* and *j* represented the individual and municipality, respectively.

However, relative income (*RI*) was expected to be highly correlated with equivalized income (*EI*). To avoid multicollinearity, we introduced RI dummy variable (*d_RI*), in addition to the continuous RI variable.

Therefore, we captured relative income using a dummy valuable (*d_RI*) as follows:

$$LS_i^* = \alpha + \sum_{m=1}^{l} \beta_k X_i^m + \delta EI_i + \pi d\_RI_{ij} + \varphi W_{ij} + u_i \tag{2}$$

[8]As shown in Table 1, an ordered logit/probit estimation was also used in Shields et al. (2009), Helliwell and Huang (2005), and Blanchflower and Oswald (2004). Where a dependent variable has more than two categories and the values of each category have a meaningful sequential order where a value is indeed higher than the previous one, an ordinal logit/probit model should be adopted.

[9]For the ordered logit model, *u* is logistically distributed with $F(z) = e^z/(1 + e^z)$.

3.2 Data

The sample included 10,416 respondents living in 1,162 municipalities,[10] distributed across the nation.

3.2.1 Dependent Variable: LS

We assessed subjective LS based on answers to the following question: "How satisfied are you currently with life overall?" because this is the most commonly used measure of LS. In the survey, this variable is scored on an 11-point scale, from 0 (very unsatisfied) to 10 (very satisfied). Nearly 35% of the respondents reported their LS to be 7 or 8 (Table 4).

3.2.2 Independent Variables

The independent variables were individual attributes, municipality size, and individual equivalized income and relative income.

(1) Individual Attributes:

Individual attributes included gender, age, marital status, housing status, educational attainments, and occupational status. The data cover male and female residents in Japan between 20 and 69 years old. The mean age was about 46 years old. Marital status was categorized into married, unmarried and divorced (including widowed). In all, 64.4% of the respondents were married and 67.9% of the respondents were homeowners. Educational attainment was categorized into junior high school or high school graduate, junior college or vocational school degree, and university or post-graduate degree. Respondents who had junior high school or high school graduate and junior college or vocational school education were slightly more than 20% respectively. About 54% of respondents had university or post-graduate degrees.[11]

Occupational statuses were categorized into regular employees, non-regular employees, self-employed and family employees, unemployed, and non-workers (the retired, homemakers, and students). The percentages for these five categories were 42.2%, 19.5%, 9.0%, 2.4%, and 26.9%, respectively. Regular employees included company executives, permanent employees, and civil servants, whereas non-regular

[10] In 2015, there were 1,741 municipalities in Japan. Our sample covered 66.7% of the municipalities.

[11] We regarded educational attainment as a proxy for bridging social capital, because individuals with undergraduate or postgraduate university degrees usually have more opportunities to meet new people and to build bridging social capital than those who do not pursue university education.

employees were part-time workers, temporary staff, contract workers, and fixed-term staff.

(2) Municipal-Level Variable:

The sizes of the municipalities were controlled using the regional-level variable (towns and villages, small cities [population < 200,000], medium-sized cities [population 200,000–700,000], and the 20 major cities [population > 700,000]).[12]

(3) Absolute Income and Relative Income:

In our survey, the absolute income was defined by the equivalized income; relative income was calculated by dividing the equivalized income by the average municipal taxable earned income. As noted above, we additionally adopted two RI dummy variables (*d_RI*) instead of the continuous relative income variable (*RI*).

One variable was the quartile dummy variable. The RI dummy variables for the quartiles were "Low level," "Lower middle level," "Upper middle level," and "High level." We adopted "Lower middle" as a reference. The other was a categorical RI dummy that consisted of four categories ($RI \leq 0.5$, $0.5 < RI \leq 2$, $2 < RI \leq 4$, and $4 < RI$), with reference to Shields et al. (2009). $RI \leq 0.5$ included respondents who earned 50% of municipal average income or less, $0.5 < RI \leq 2$ included respondents who earned greater than or equal to 50% and less than or equal to 200% of municipal average income, $2 < RI \leq 4$ included respondents who earned greater than or equal to 200% and less than or equal to 400% of municipal average income, and $4 < RI$ included respondents who earned more than 400% of municipal average income. We adopted $0.5 < RI \leq 2$ as a reference.

Table 5 presents the descriptive statistics for LS, individual-level variables, and regional-level variables.

4 Empirical Results

We performed three ordered logit regressions with LS as the dependent variable. All three models included all individual and regional variables, so that we could isolate the association between LS and relative income after controlling for all variables, as is shown in Eqs. (1) and (2). The continuous relative income variable RI was adopted for Model [A], while quartile RI dummy and categorical RI dummy were adopted in Model [B] and Model [C], respectively. Table 6 shows the estimated results.[13]

(1) Individual Attributes:

The results obtained on individual attributes were generally as expected, indicating that LS was affected by certain individual attributes. Some examples follow. Age

[12]These are formally termed government-designated cities; these require the population to be over 500,000. All 20 such cities had populations over 700,000 in 2015.

[13]The null hypothesis of the log likelihood ratio chi-square (LR χ^2) test that all predictor variables do not contribute to the model was rejected in Model [A]–[C].

Table 5 Descriptive statistics

		Mean	Std. Dev.	Min	Max
Dependent variable					
Life satisfaction: current life overall					
0 (=very unsatisfied)–10 (=very satisfied)		5.922	2.383	0	10
Independent variables					
Individual attributes					
Age	20s [Reference]	0.147	0.354	0	1
	30s	0.218	0.413	0	1
	40s	0.208	0.406	0	1
	50s	0.202	0.401	0	1
	60s	0.226	0.418	0	1
Gender	Female [Reference: Male]	0.486	0.500	0	1
Marital status	Married [Reference]	0.646	0.478	0	1
	Unmarried	0.291	0.454	0	1
	Divorced	0.062	0.242	0	1
Home ownership	Homeowner [Reference: Renter]	0.679	0.467	0	1
Educational attainment	Junior or high school graduate [Reference]	0.250	0.433	0	1
	Junior college or vocational school degree	0.216	0.411	0	1
	University undergraduate or postgraduate degree	0.535	0.499	0	1
Occupational status	Regular employee [Reference]	0.423	0.494	0	1
	Non-regular employee	0.195	0.396	0	1
	Unemployed	0.024	0.153	0	1
	Retired and homemakers	0.269	0.444	0	1
	Self-employed and family employee	0.090	0.286	0	1
Municipality size	Towns and villages [Reference]	0.082	0.275	0	1
	Small cities (population < 200,000)	0.373	0.484	0	1
	Medium-sized cities (population 200,000–700,000)	0.251	0.433	0	1
	20 major cities (population > 700,000)	0.294	0.456	0	1

(continued)

Table 5 (continued)

		Mean	Std. Dev.	Min	Max
Absolute income					
Equivalized income (EI): (million yen)		4.071	2.868	0.177	32.600
Relative income					
Relative income (RI)	(=EI/Municipal average taxable income)	2.811	1.978	0.093	36.240
RI dummy	Low RI (1st-quartile)	0.250	0.433	0	1
	Lower middle RI (2nd-quartile) [Reference]	0.250	0.433	0	1
	Upper middle RI (3rd-quartile)	0.250	0.433	0	1
	High RI (4th-quartile)	0.250	0.433	0	1
RI Dummy	RI ≤ 0.5 (Y ≤ 0.5Y*)	0.030	0.170	0	1
Y: own income	0.5 < RI ≤ 2 (0.5Y* < Y≤ 2Y*) [Reference]	0.344	0.475	0	1
Y*: reference income	2 < RI ≤ 4 (2Y* < Y≤ 4Y*)	0.451	0.498	0	1
	4 < RI (Y > 4Y*)	0.175	0.380	0	1

and LS showed a U-shape association. The LS of individuals in their 30s–50s was rather low. Women reported higher satisfaction than men. The unmarried and the divorced reported lower satisfaction than the married. Homeowners reported higher satisfaction than renters. The retired and homemakers reported higher satisfaction and non-regular employees and the unemployed reported lower satisfaction than regular employees. Individuals from the groups having junior college or vocational school degrees and having undergraduate or postgraduate degrees reported higher satisfaction than other groups.[14]

(2) Municipal-Level Variable:

There was no association between municipality size and LS in any model, except that there was a significantly positive association between the 20 major cities and LS at the 10% level in Model [B].

(3) Absolute Income versus Relative Income:

Regression analysis showed that there was a significant positive association between equivalized income, i.e., absolute income, and LS in all models. In short, individuals with higher incomes enjoyed higher satisfaction.

[14]Individuals with undergraduate or postgraduate degrees are expected to have higher levels of bridging social capital than others. The relationships between life satisfaction and educational background revealed here are supported by Putnam (1993, 2000, 2015).

Table 6 Ordered logit regression estimation results

Variables	Model	[A]	[B]	[C]
		coef.	coef.	coef.
		[z-value]	[z-value]	[z-value]
Individual attributes				
Age [Reference: 20s]	30s	−0.339***	−0.344***	−0.349***
		[−5.541]	[−5.624]	[−5.695]
	40s	−0.524***	−0.535***	−0.547***
		[−8.085]	[−8.249]	[−8.413]
	50s	−0.619***	−0.628***	−0.634***
		[−9.096]	[−9.200]	[−9.303]
	60s	−0.112	−0.104	−0.118*
		[−1.610]	[−1.494]	[−1.702]
Gender [Reference: Male]	Female	0.439***	0.425***	0.428***
		[10.947]	[10.588]	[10.653]
Marital status [Reference: Married]	Divorced	−0.628***	−0.584***	−0.596***
		[−8.415]	[−7.788]	[−7.961]
	Unmarried	−1.015***	−0.981***	−0.988***
		[−22.268]	[−21.403]	[−21.583]
Home ownership [Reference: Renter]	Homeowner	0.305***	0.293***	0.302***
		[7.687]	[7.375]	[7.604]
Educational attainment [Reference: Junior or high school graduate]	Junior college or vocational school degree	0.150***	0.145***	0.142***
		[2.898]	[2.791]	[2.740]

(continued)

Table 6 (continued)

Variables	Model	[A]	[B]	[C]
		coef.	coef.	coef.
		[z-value]	[z-value]	[z-value]
	University undergraduate or postgraduate degree	0.289***	0.282***	0.283***
		[6.523]	[6.385]	[6.393]
Occupational status [Reference: Regular employee]	Non-regular employee	−0.265***	−0.205***	−0.225***
		[−5.093]	[−3.918]	[−4.303]
	Unemployed	−1.081***	−0.963***	−0.988***
		[−9.228]	[−8.174]	[−8.375]
	Retired and homemakers	0.150***	0.226***	0.213***
		[2.909]	[4.350]	[4.091]
	Self-employed and family employee	−0.129*	−0.065	−0.086
		[−1.950]	[−0.982]	[−1.294]
Municipality size				
Small cities [Reference: Towns and villages]	(population < 200,000)	0.081	0.106	0.100
		[1.213]	[1.583]	[1.494]
Medium-sized cities	(population 200,000–700,000)	0.023	0.065	0.056
		[0.330]	[0.921]	[0.799]
20 major cities	(population > 700,000)	0.041	0.117*	0.099
		[0.567]	[1.655]	[1.411]
Absolute income				
Equivalized income		0.090***	0.066***	0.073***
		[5.694]	[7.120]	[7.531]

(continued)

Table 6 (continued)

Variables	Model	[A]	[B]	[C]
		coef.	coef.	coef.
		[z-value]	[z-value]	[z-value]
Relative income (equivalized income/municipal average taxable income)				
Relative income		0.042*		
		[1.839]		
Low RI dummy [Reference: 2nd-quartile]	(1st-quartile)		−0.371***	
			[−7.171]	
Upper middle RI dummy	(3rd-quartile)		0.134***	
			[2.690]	
High RI dummy	(4th-quartile)		0.246***	
			[3.941]	
RI ≤ 0.5 [Reference: 0.5 < RI ≤ 2]	(Y ≤ 0.5Y*)			−0.239**
				[−2.126]
2 < RI ≤ 4	(2Y* < Y≤4Y*)			0.345***
				[7.790]
4 < RI	(Y > 4Y*)			0.451***
				[5.873]
Constant				
K1		−3.216***	−3.375***	−3.140***
K2		−2.738***	−2.895***	−2.660***
K3		−2.065***	−2.218***	−1.984***

(continued)

Table 6 (continued)

Variables	Model [A]	[B]	[C]
	coef.	coef.	coef.
	[z-value]	[z-value]	[z-value]
K4	−1.352***	−1.500***	−1.267***
K5	−0.858***	−1.001***	−0.770***
K6	0.010	−0.127	0.103
K7	0.588***	0.455***	0.684***
K8	1.440***	1.311***	1.538***
K9	2.682***	2.553***	2.781***
K10	3.638***	3.508***	3.735***
N	10,416	10,416	10,416
LR χ^2	1,693	1,785	1,759
Log-likelihood	−22,061	−22,015	−22,027

Note $^{*}p < 0.10$, $^{**}p < 0.05$, $^{***}p < 0.01$

Regarding relative income, we found a positive but no association between relative income and LS in Model [A]. The association might be weak, because there was multi-collinearity between household earned income and relative income. On the other hand, empirical results from Model [B] indicated that there is a significant negative association between the Low RI dummy and LS at the 1% level and a significant positive association between both the Upper middle RI dummy and the High RI dummy and LS at the 1% level. The empirical results from Model [C] also indicated that there was a negative association between the RI $\leq$ 0.5 dummy and LS at the 5% level, and there were significant positive associations between both 2 < RI $\leq$ 4 and the 4 < RI dummy and LS at the 1% level. These results support the relative income hypothesis.

5 Marginal Effects

5.1 Average Discrete Change

To evaluate the effects of RI on LS, we obtained marginal effects from an ordered logit estimation result. A marginal effect measures the change in the probability of an outcome m for a change in independent variable x_l. There are two types of marginal effects. One is a marginal change which indicates the effect of an instantaneous change in x_l, the other is a discrete change which indicates the effect of a discrete change in x_l.

For the binary independent variable, reasonable change was from 0 to 1. According to Long and Freese (2014), discrete change is the change in the probability of m for change in x_l from the start value to the end value, holding all other variables constant. All independent variables except equivalized income were factor variables. The discrete change in the categorical variable was the change in the probability of m for change in x_l from 0 to 1, holding all other variables constant. On the other hand, for continuous independent variables such as equivalized income, which is measured in terms of 10,000 yen in our sample, because the marginal change in income is the instantaneous rate of change, this did not exactly approximate the effects of unit changes in income on the probability of outcomes. A discrete change is the change in the probability of m for a change in standard deviation and any amount (such as 1 million yen) in x_l, holding all other variables constant. Thus, it was plausible for us to adopt discrete change.

The average discrete change (ADC) is the mean of the marginal change calculated at the observed values for all observations in the sample, as in Eq. 3.

$$\mathrm{ADC} = \frac{1}{N}\sum_{i=1}^{N}\frac{\Delta \Pr(LS_i = m|\mathrm{X} = x_i)}{\Delta x_l} \tag{3}$$

here, m = 0, …,10. ADC implies, on average, that a discrete change in x_l increases the probability of m by y percentage points.[15] The ADCs for all independent variables are shown in Table 7.

5.2 Evaluation of Marginal Effect of Relative Income

We have found marginal effects for both discrete and continuous independent variables in our results. First, we found that discrete changes in age, gender, marital status, and occupational status had relatively larger marginal effects of predicted probability. As shown in Fig. 1, discrete changes in age (such as from 20s to 30s, from 30s to 40s, from 40s to 50s, and from 50s to 60s) changed the probability from −3.1 percentage points to 4.7 percentage points on average. For example, on average, the discrete change from 20s to 30s was significantly associated with a −3.1 percentage points change in the probability of choosing outcome 8 (LS = 8), while the discrete change from the 50s to 60s was associated with a 4.7 percentage points increase in the probability of choosing outcome 8 (LS = 8).

Figure 2 shows that being a woman was also associated with a 4.0 percentage points increase in the probability of choosing outcome 8 (LS = 8) and a 2.2 percentage points decrease in the probability of choosing outcome 5 (LS = 5) on average. Discrete changes in marital status (such as Unmarried vs. Married, Divorced vs. Married and Divorced vs. Unmarried) changed the probability from −9.3 percentage points to 5.3 percentage points on average. In fact, on average, being unmarried was associated with a 5.3 percentage points increase in the probability of choosing outcome 3 (LS = 3) and a 9.3 percentage points decrease the probability of choosing outcome 8 (LS = 8) relative to a married respondent. Figure 3 shows that discrete changes in occupational status (such as Unemployed vs. Regular employee) changed probabilities by −8.4 percentage points to 5.2 percentage points on average. For example, on average, unemployed was associated with an 8.4 percentage points decrease in the probability of choosing outcome 8 (LS = 8), while it increased the probability of choosing outcome 3 (LS = 3) by 5.2 percentage points, relative to regular employees. However, non-regular employees and the retired showed significantly smaller marginal effects of predicted probability than the unemployed. There was no statistically significant difference between the self-employed and regular employees.

Second, unlike these factors, discrete changes in educational attainment and home ownership had relatively smaller marginal effects on predicted probability, as shown in Fig. 4.

---

[15]Another measure of marginal effect is a discrete change at the mean (MDC). The MDC was computed for all variables held at their means. MDC = $\Delta \Pr(LS_i = m|\overline{X}, x_l = \overline{x_l})/\Delta x_l$. To check the robustness of this analysis, we also calculated MDC. We found that ADC and MDC showed a similar tendency, as shown in the Appendix Table.

Table 7 Marginal effects: average discrete change (ADC) in independent variables (Unit: %)

Life satisfaction	0	1	2	3	4	5	6	7	8	9	10
Age											
30s versus 20s	0.9***	0.5***	1.0***	1.6***	1.3***	1.9***	0.5***	−0.8***	−3.1***	−1.9***	−1.8***
40s versus 30s	0.7***	0.3***	0.7***	1.0***	0.7***	0.9***	0.1**	−0.8***	−1.8***	−0.9***	−0.8***
50s versus 40s	0.3	0.2	0.3	0.4	0.3	0.3	0.0	−0.4	−0.8	−0.4	−0.3
60s versus 50s	−1.7***	−0.8***	−1.7***	−2.5***	−1.8***	−2.5***	−0.3***	1.8***	4.7***	2.6***	2.3***
60s versus 20s	0.3*	0.2*	0.3*	0.5*	0.4*	0.7*	0.2*	−0.2*	−1.0*	−0.7*	−0.7*
Gender											
Female versus male	−1.3***	−0.7***	−1.4***	−2.0***	−1.6***	−2.2***	−0.3***	1.5***	4.0***	2.2***	1.9***
Marital status											
Divorced versus married	1.7***	0.9***	1.9***	3.0***	2.4***	3.4***	0.5***	−2.4***	−5.8***	−3.0***	−2.5***
Unmarried versus married	3.3***	1.7***	3.6***	5.3***	3.9***	4.6***	−0.2**	−4.9***	−9.3***	−4.4***	−3.6***
Unmarried versus divorced	1.7***	0.8***	1.7***	2.3***	1.5***	1.2***	−0.7***	−2.5***	−3.5***	−1.4***	−1.1***
Home ownership											
Owner versus renter	−1.0***	−0.5***	−1.0***	−1.5***	−1.1***	−1.5***	−0.1***	1.2***	2.8***	1.5***	1.3***
Educational attainment											
Junior college versus high school	−0.5***	−0.2***	−0.5***	−0.7***	−0.5***	−0.6***	0.0	0.6***	1.3***	0.7***	0.6***
University versus junior college	−0.4***	−0.2***	−0.5***	−0.7***	−0.5***	−0.7***	−0.1***	0.5***	1.3***	0.7***	0.6***

(continued)

Table 7 (continued)

Life satisfaction	0	1	2	3	4	5	6	7	8	9	10
Occupational status											
Non regular versus regular	0.7***	0.4***	0.8***	1.1***	0.8***	1.1***	0.1**	−1.0***	−2.1***	−1.1***	−0.9***
Unemployed versus regular	4.6***	2.2***	4.2***	5.2***	3.2***	2.5***	−1.5***	−5.5***	−8.4***	−3.7***	−2.9***
Retired versus regular	−0.6***	−0.3***	−0.7***	−1.0***	−0.8***	−1.2***	−0.3***	0.6***	2.0***	1.1***	1.0***
Self employed versus regular	0.3	0.1	0.3	0.4	0.3	0.4	0.1	−0.3	−0.8	−0.4	−0.4
Absolute income (equivalized income)											
⊿1 Standard deviation	−0.6***	−0.3***	−0.6***	−1.0***	−0.8***	−1.1***	−0.3***	0.6***	1.9***	1.1***	1.0***
⊿0.5 Millon	−0.1***	−0.1***	−0.1***	−0.2***	−0.1***	−0.2***	0.0***	0.1***	0.3***	0.2***	0.2***
⊿1 Million Yen	−0.2***	−0.1***	−0.2***	−0.3***	−0.3***	−0.4***	−0.1***	0.2***	0.7***	0.4***	0.3***
⊿2 Million Yen	−0.4***	−0.2***	−0.5***	−0.7***	−0.5***	−0.8***	−0.2***	0.4***	1.3***	0.8***	0.7***
⊿3 Million Yen	−0.6***	−0.3***	−0.7***	−1.0***	−0.8***	−1.2***	−0.3***	0.6***	2.0***	1.2***	1.1***
⊿4 Million Yen	−0.8***	−0.4***	−0.9***	−1.3***	−1.1***	−1.6***	−0.4***	0.8***	2.6***	1.6***	1.5***
⊿5 Million Yen	−1.0***	−0.5***	−1.1***	−1.6***	−1.3***	−2.0***	−0.6***	0.9***	3.3***	2.0***	1.9***
Relative income											
$0.5 < RI \leq 2$ versus $RI \leq 0.5$	−1.0*	−0.5**	−1.0**	−1.3**	−0.9**	−0.8***	0.2	1.3**	2.2**	1.0**	0.8**
$2 < RI \leq 4$ versus $0.5 < RI \leq 2$	−1.1***	−0.6***	−1.2***	−1.7***	−1.3***	−1.7***	−0.2***	1.4***	3.2***	1.7***	1.4***
$RI > 4$ versus $2 < RI \leq 4$	−0.3*	−0.1*	−0.3*	−0.5*	−0.4*	−0.6*	−0.2	0.3*	1.0*	0.6*	0.5*

Note *$p < 0.10$, **$p < 0.05$, ***$p < 0.01$

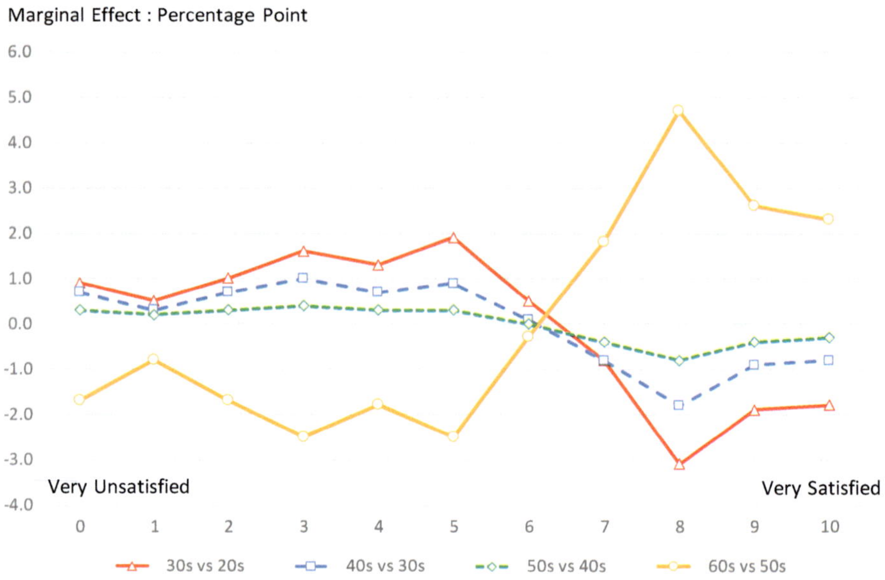

Fig. 1 Marginal effects: average discrete changes for changes in age

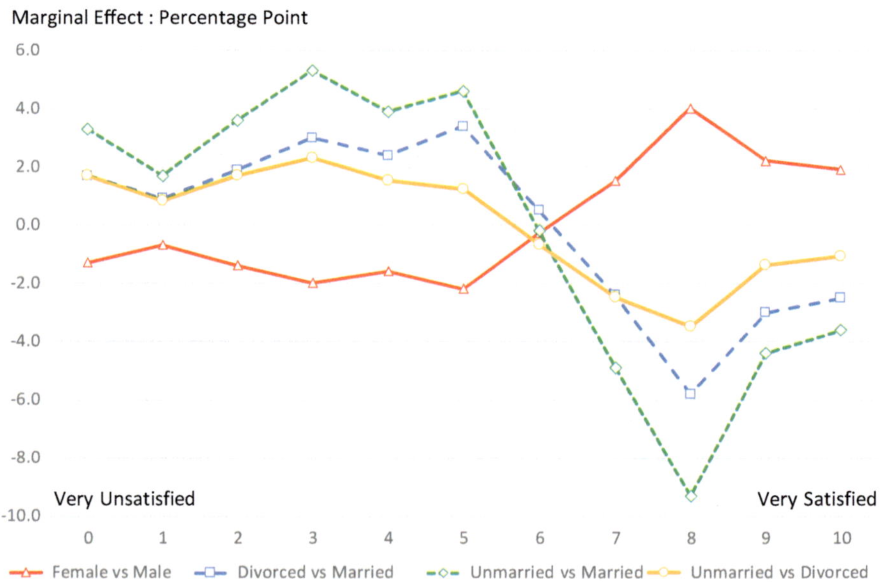

Fig. 2 Marginal effects: average discrete changes in gender and marital status

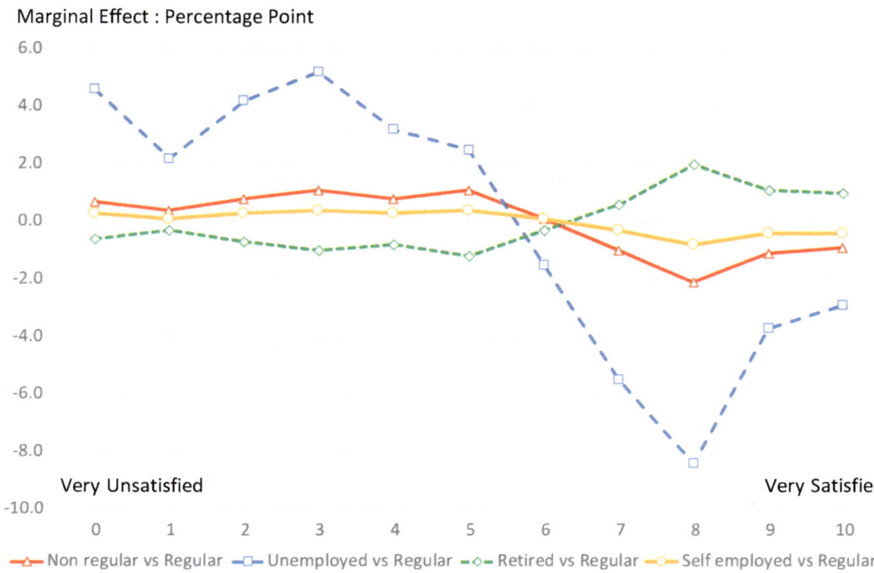

Fig. 3 Marginal effects: average discrete changes in occupational status

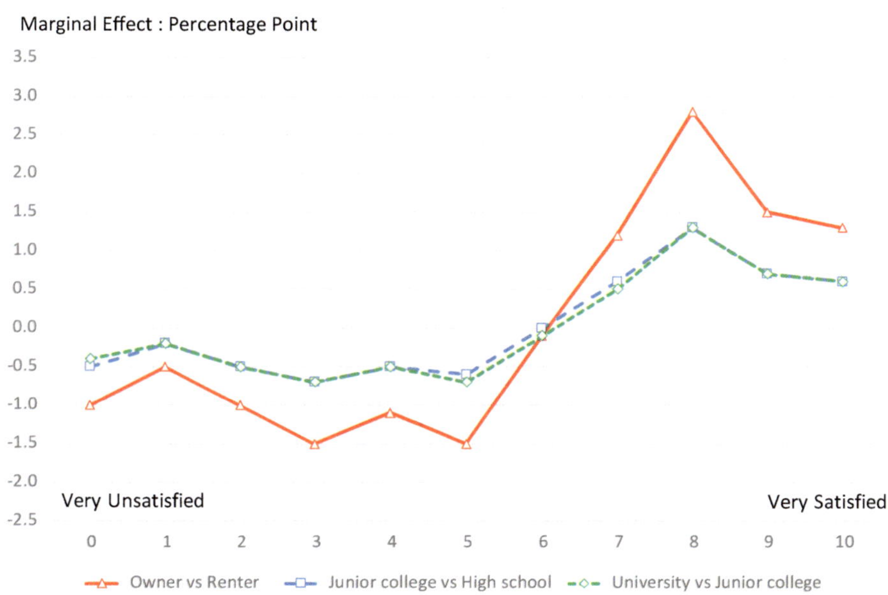

Fig. 4 Marginal effects: average discrete changes in home ownership and educational attainment

Finally, Table 7 shows that discrete changes in both absolute and relative incomes had relatively smaller marginal effects on predicted probability than discrete changes in individual attributes such as age, marital status, gender, and occupational status. A standard-deviation increase in equivalized income (about 2.9 million yen) changed the probability from -1.1 percentage points to 1.9 percentage points on average. A standard-deviation increase in equivalized income decreased the probability of choosing outcome 5 ($LS = 5$) by 1.1 percentage points and increased the probability of choosing outcome 8 ($LS = 8$) by 1.9 percentage points.[16]

With respect to relative income, an increase from $0.5 \leq RI$ to $0.5 < RI \leq 2$ significantly decreased the probability of choosing outcome 3 ($LS = 3$) by 1.3 percentage points and increased the probability of choosing outcome 8 ($LS = 8$) by 2.2 percentage points on average. In addition, an increase from $0.5 < RI \leq 2$ to $2 < RI \leq 4$ significantly decreased the probability of choosing outcome 3 ($LS = 3$) by 1.7 percentage points and increased the probability of choosing outcome 8 ($LS = 8$) by 3.2 percentage points on average. By contrast, Fig. 5 and Table 7 show that an increase from $2 < RI \leq 4$ to $4 < RI$ had slight differences in marginal effects. These results indicate that on average, (1) when own income was less than double

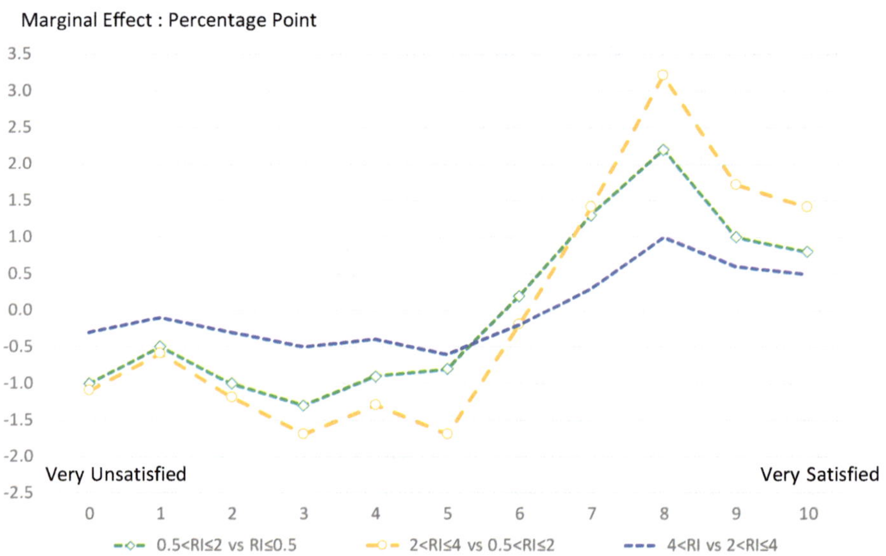

Fig. 5 Marginal effects: average discrete changes in relative income

[16]For reference, as shown in Table 7 that, on average, a 1-million-yen change in equivalized income decreases the probability of choosing outcome 5 ($LS = 5$) by 0.4 percentage points and increases the probability of choosing outcome 8 ($LS = 8$) by 0.7 percentage points. On average, a 3-million-yen change in equivalized income decrease the probability of choosing outcome 5 ($LS = 5$) by 1.2 percentage points and increases the probability of choosing outcome 8 ($LS = 8$) by 2.2 percentage points. It is clear that standard-deviation increases in equivalized income (about 2.9 million yen) and a 3-million-yen changes have almost same marginal effect on predicted probability.

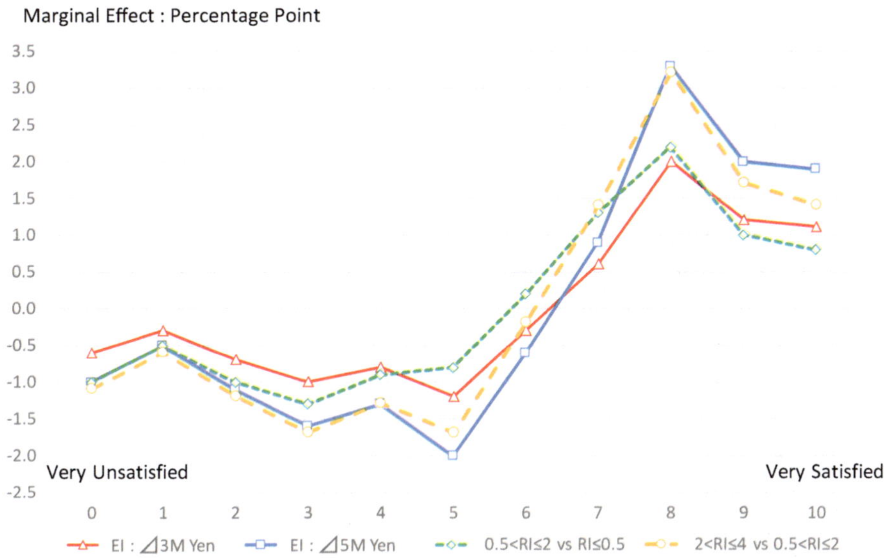

Fig. 6 Marginal effects: average discrete changes in absolute income and relative income

the reference income, the marginal effect of predicted probability of relative income is relatively small; (2) when own income was twice more than the reference income, the marginal effect of predicted probability of relative income is a considerably large; and (3) when own income was four times the reference income or higher, its marginal effect of predicted probability only slightly increased even if relative income additionally escalates.

In addition, Fig. 6 and Table 7 show that he marginal effect of predicted probability of a change from $0.5 < RI \leq 2$ to $2 < RI \leq 4$ and about a 3-million-yen increase in absolute income were equivalent on average and the marginal effect of the predicted probability of a change from $2 < RI \leq 4$ to $4 < RI$ and about a 5-million-yen increase in absolute income were equivalent on average.

We also found that the marginal effect of the predicted probability of these income changes were approximately equal to the difference between being retired or a home-maker and being a regular employee and between being a homeowner and being a renter, on average, as shown in Table 7.

6 Conclusions

In this study, we examined the relative income hypothesis, using Japanese nationwide survey data. The main findings of the study were as follows:

First, we found that not only absolute but also relative income had a significant positive association with LS. Therefore, our empirical results provide support for both income hypotheses: individual utility depends on not only on one's own income but also on one's income relative to those of others in area where individual has his or her residence.

Second, it was shown that with respect to marginal effects, discrete changes in both absolute and relative income had relatively smaller marginal effects on predicted probability than those of discrete changes in individual attributes such as age, marital status, and occupational status, on average.

Third, with respect to relative income, a change from $0.5 < RI \leq 2$ to $2 < RI \leq 4$ increased the probability of choosing higher satisfaction largely; marginal effect of the predicted probability of this change and of an about 3-million-yen increase in absolute income were almost equivalent.

Our contribution to the study of happiness and well-being is that our results support the relative income hypothesis in Japan: individual life satisfaction depends not only on one's own income but also on one's income relative to average income of the municipality where an individual has his or her residence. In addition, we found diminishing marginal utility with relative income. When relative income was sufficiently large (i.e., more than four times the reference income), its marginal effect of predicted probability only slightly increased, even if relative income additionally increases.

Further research remains to be done. Our results indicate that individual LS greatly depends on the individual's life stage. For example, change from one's 20s to one's 30s reduces the probability of getting higher LS, while the change from one's 50s to one's 60s increases it, on average. This may be because those in their 20s may envision a bright future for themselves and look positively at their current status, while those in their 30s and 40s have faced the severe reality of employment and labor market. Meanwhile, the elderly Japanese were found to have higher LS, although as a group they don't have a bright prospect for their future as younger generation. These results might indicate that the determinants of individual's happiness differ among age groups. For example, the working generation place a high value on their career success, whereas the elderly generation would not place a high value on it. As people retire from work and become aged, they turn their interests toward social ties between familiar people, such as family, friends, and neighborhood. However, further research is needed to understand this relationship between individual's happiness and age; this issue must be addressed in the future.

Acknowledgements This work was supported by the MEXT-Supported Program for the Strategic Research Foundation at Private Universities of Japan, 2014–2018 (S1491003). "International Comparative Surveys on Lifestyle and Values" were designed and conducted by the Center for Social Well-being Studies, Institute for the Development of Social Intelligence, Senshu University, Japan, in collaboration with Social Well-being Research Consortium in Asia.

Appendix

Marginal effects: discrete change in independent variables at the mean (MDC) (Unit: %)

Life satisfaction	0	1	2	3	4	5	6	7	8	9	10
Age											
30s versus 20s	0.7***	0.4***	1.0***	1.7***	1.6***	2.6***	0.7***	−1.4***	−3.9***	−1.9***	−1.5***
40s versus 30s	0.5***	0.3***	0.7***	1.1***	0.9***	1.3***	0.0	−1.2***	−2.1***	−0.9***	−0.7***
50s versus 40s	0.3	0.2	0.3	0.5	0.4	0.5	−0.1	−0.6	−0.9	−0.4	−0.3
60s versus 50s	−1.3***	−0.7***	−1.7***	−2.8***	−2.4***	−3.5***	−0.3***	2.8***	5.6***	2.4***	1.9***
60s versus 20s	0.2*	0.1*	0.3*	0.5*	0.5*	0.9*	0.3*	−0.4*	−1.3*	−0.7*	−0.6*
Gender											
Female versus male	−1.1***	−0.6***	−1.4***	−2.3***	−2.0	−3.0***	−0.4***	2.3***	4.7***	2.1***	1.6***
Martial status											
Divorced versus married	1.4***	0.8***	1.9***	3.1***	2.7***	4.1***	0.6***	−3.0***	−6.5***	−2.9***	−2.3***
Unmarried versus married	3.0***	1.6***	3.6***	5.6***	4.5***	5.6***	−0.3**	−6.0***	−10.1***	−4.3***	−3.2***
Unmarried versus divorced	1.5***	0.8***	1.7***	2.5***	1.8***	1.4***	−0.9**	−3.0***	−3.6***	−1.3***	−0.9***
Home ownership											
Owner versus renter	−0.8***	−0.4***	−1.0***	−1.6***	−1.4***	−2.0***	−0.1***	1.7***	3.2***	1.4***	1.1***
Education attainment											
Junior college versus high school	−0.4***	−0.2***	−0.5***	−0.8***	−0.7***	−0.9***	0.0	0.9***	1.5***	0.6***	0.5***
University versus junior college	−0.3***	−0.2***	−0.4***	−0.7***	−0.6***	−1.0***	−0.1***	0.7***	1.5***	0.7***	0.5***
Occupational status											

(continued)

(continued)

Life satisfaction	0	1	2	3	4	5	6	7	8	9	10
Non regular versus regular	0.6***	0.3***	0.8***	1.3***	1.1***	1.5***	0.0	−1.3***	−2.4***	−1.0***	−0.8***
Unemployed versus regular	3.9***	2.1***	4.4***	6.2***	4.3***	3.3***	−2.2***	−7.2***	−9.0***	−3.3***	−2.4***
Retired versus regular	−0.5***	−0.3***	−0.6***	−1.1***	−1.0***	−1.6***	−0.4***	0.9***	2.4***	1.1***	0.9***
Self employed versus regular	0.2	0.1	0.3	0.5	0.4	0.6	0.1	−0.5	−0.9	−0.4	−0.3
Absolute income (equivalized income)											
⊿1 Standard Deviation	−0.5***	−0.3***	−0.6***	−1.1***	−1.0***	−1.6***	−0.3***	1.0***	2.3***	1.1***	0.9***
⊿0.5 Millon	−0.1***	−0.1***	−0.1***	−0.2***	−0.2***	−0.3***	0.0**	0.2***	0.4***	0.2***	0.1***
⊿1 Million Yen	−0.2***	−0.1***	−0.2***	−0.4***	−0.3***	−0.5***	−0.1**	0.4***	0.8***	0.4***	0.3***
⊿2 Million Yen	−0.3***	−0.2***	−0.4***	−0.7***	−0.7***	−1.1***	−0.2**	0.7***	1.6***	0.7***	0.6***
⊿3 Million Yen	−0.5***	−0.3***	−0.6***	−1.1***	−1.0***	−1.6***	−0.3**	1.0***	2.4***	1.1***	0.9***
⊿4 Million Yen	−0.6***	−0.4***	−0.8***	−1.4***	−1.3***	−2.2***	−0.5**	1.2***	3.3***	1.6***	1.2***
⊿5 Million Yen	−0.8***	−0.4***	−1.0***	−1.8***	−1.6***	−2.8***	−0.7**	1.4***	4.1***	2.0***	1.6***
Relative income											
0.5 < RI ≤ 2 versus RI ≤ 0.5	−0.8*	−0.4*	−1.0**	−1.5**	−1.1**	−1.1***	0.4	1.7**	2.3**	0.9**	0.6**
2 < RI ≤ 4 versus 0.5 < RI ≤ 2	−0.9***	−0.5***	−1.1***	−1.9***	−1.6***	−2.3***	−0.2***	2.0***	3.7***	1.6***	1.2***
RI > 4 versus 2 < RI ≤ 4	−0.2*	−0.1*	−0.3*	−0.5*	−0.5*	−0.8*	−0.2	0.4*	1.2*	0.6*	0.5*

Note *$p < 0.10$, **$p < 0.05$, ***$p < 0.01$

References

Blanchflower, D. G., & Oswald, A. J. (2004). Well-being over time in Britain and the USA. *Journal of Public Economics, 88,* 1359–1386.

Clark, A. E., & Oswald, A. J. (1994). Unhappiness and unemployment. *Economic Journal, 104*(424), 648–659.

Clark, A. E., Westergård-Nielsen, N., & Kristensen, N. (2009). Economic satisfaction and income rank in small neighborhoods. *Journal of European Economic Association, 7*(2–3), 519–527.

Duesenberry, J. S. (1949). *Income, saving and the theory of consumer behavior*. MA: Harvard University Press.

Easterlin, R. (1974). Does economic growth improve the human lot?: Some empirical evidence. In R. David & M. Reder (Eds.), *Nations and households in economic growth: Essays in honor of Moses Abramowitz* (pp. 89–125). UT: Academic Press.

Frey, B., & Stutzer, A. (2002). What can economists learn from happiness research? *Journal of Economic Literature, 40*(2), 402–435.

Friedman, M. (1957). *A theory of the consumption function*. NJ: Princeton University Press.

Harada, H. (2016a). The significance and availability of happiness study. *The Senshu Social Well-being Review, 2,* 9–39.

Harada, H. (2016b). Happiness in Japan: From the viewpoint of age, sex and relative wealthiness. *The Senshu Social Well-being Review, 3,* 1–17.

Helliwell, J. F., & Huang, H. (2005). *How's the job? Well-being and social capital in the workplace* (NBER Working Paper No. 11759).

Inaba, Y., Wada, Y., Ichida, Y., & Nishikawa, M. (2015). Which part of community social capital is related to life satisfaction and self-rated health? A multilevel analysis based on a nationwide mail survey in Japan. *Social Science and Medicine, 142,* 169–182.

Long, J. S., & Freese, J. (2014). *Regression models for categorical dependent variables using stata* (3rd ed.). TX: Stata Press.

Luttmer, E. F. (2005). Neighbours as negatives: Relative earnings and well-being. *Quarterly Journal of Economics, 120*(3), 963–1002.

Mangyo, E., & Park, A. (2011). Relative deprivation and health which reference groups matter? *Journal of Human Recourses, 46*(3), 459–481.

Oshio, T., & Kobayashi, M. (2009). Income inequality, area-level poverty, perceived aversion to inequality, and self-rated health in Japan. *Social Science and Medicine, 69,* 317–326.

Oshio, T., & Kobayashi, M. (2011). Area-level income inequality and individual happiness: Evidence from Japan. *Journal of Happiness Studies, 12*(4), 633–649.

Putnam, R. (1993). *Making democracy work: Civic tradition in modern Italy*. NJ: Princeton University Press.

Putnam, R. (2000). *Bowling alone: The collapse and revival of American community*. NY: Simon and Schuster.

Putnam, R. (2015). *Our kids: The American dream in crisis*. NY: Simon and Schuster.

Shields, M. A., Price, S. W., & Wooden, M. (2009). Life satisfaction and the economic and social characteristics of neighbourhoods. *Journal of Population Economics, 22,* 421–443.

Stiglitz, J. E., Sen, A., & Fitoussi, J. (2010). *Mis-measuring our lives: Why GDP doesn't add up?*. NY: The New Press.

Sumi, E. (2016). Subjective well-being and income inequality. *The Senshu Social Well-being Review, 2,* 85–98.

Hiroo Harada is Professor Emeritus at Senshu University in Japan from April 2019, and was Professor of Economics at Senshu University until March 2019. He is President of the Association for the Study of Political Society, Kanto, Japan, and Councilor of the Japan Association for Planning and Public Management. He was President of Japan Public Choice Society, and Vice Chairman of the International Academic Society for Asian Community. He was Chair of the Center for Social Well-being Studies, 2014–2018, and the Center for Social Capital Studies, 2009–2013, both of which were at Senshu University. He is an author in *The International Encyclopedia of Political Communication* (John Wiley & Sons 2016). His articles have appeared in *The Senshu Social Well-being Review*.

Eiji Sumi is Associate Professor of Economics at Niigata University. He is Director of the Japan Institute of Public finance and Japan Association of Local Public Finance, and was former director of The Japan Public Choice Society. His current research project focuses on relative income hypothesis in comparative perspective. His articles have appeared in *The Senshu Social Well-being Review*.

Income, Intra-household Bargaining Power and the Happiness of Japanese Married Women

Xinxin Ma and Xiangdan Piao

Abstract This is an empirical study that evaluates the influence of household income, inter-household relative income, intra-household bargaining power on the happiness of Japanese married women using data from the Japanese Household Panel Survey (JHPS) conducted from 1995 and 2013. There are five major conclusions. First, the absolute income hypothesis is supported. The result is consistent with published studies for other developed and developing countries. Second, when absolute income is controlled, the inter-household relative income hypothesis is not supported: but when absolute income is not controlled, it is supported. Third, the income and education gap between a wife and her husband can negatively affect her happiness: however, the larger the amount of the husband's income controlled by the wife the greater her happiness is, and the affect is more marked for the working wife group than the non-work wife group. Fourth, the results based on the robustness checks are consistent. Fifth, other factors, such as the wife's education, and the hours of husband participation in child care or housework positively affect her happiness. The wife's age, youngest child's age, and living with parents negatively affects happiness.

Keywords Absolute income · Inter-household relative income · Intra-household bargaining power · Happiness · Japanese married women

This research was supported by JSPS (Japan Society for the Promotion Science) KAKENHI (Grant Number JP16K03611), and the project grand of Joint Usage and Research Center Program at the Institute of Economic Research, Hitotsubashi University. We are grateful to Professor Ming-Chang Tsai (Research Center for Humanities and Social Sciences, Academia Sinica, Taiwan) for his helpful suggestions. We also acknowledge the anonymous referees and editors for their many helpful comments and suggestions, and Dr. Richard Lingwood for his helpful editing work.

X. Ma (✉)
Center for Far Eastern Studies, University of Toyama, Toyama, Japan
e-mail: xxma@eco.u-toyama.ac.jp

X. Piao
Departments of Urban and Environmental Engineering, School of Engineering, Kyushu University, Kyushu, Japan
e-mail: piao.xiangdan@doc.kyushu-u.ac.jp

M.-C. Tsai and N. Iwai (eds.), *Quality of Life in Japan*, Quality of Life in Asia 13,
https://doi.org/10.1007/978-981-13-8910-8_4

According to conventional neoclassical economics, the wellbeing of the people of a country is regarded as the total value of individual utility: for the economist this poses a problem. Utility cannot be measured because individual utility is not a cardinal number but an ordinal number, therefore it is difficult to compare utility between individuals. However, since the 1980s, Happiness Economics has overturned this argument. In Happiness Economics subjective well-being (SWB) is one of the indicators reflecting the theoretical concept of individual utility. Measurement of individual utility (e.g. happiness, life satisfaction) is necessary when evaluating a social policy and recently the empirical study of happiness has become an important concern.

Two established hypotheses, the absolute income hypothesis and relative income hypothesis have been advocated to explore the impact of income on happiness (Duesenberry 1949; Leibenstein 1950).[1] Some empirical studies for Japan proofed the two hypotheses, and they indicated that both the absolute income hypothesis and relative income hypothesis are supported for Japan, but some issues remain to be analyzed. Studies based on the Collective Model proposed by Chiappori (1992), such as Browning et al. (1994), Chiappori et al. (2002), Couprie (2007), Lise and Seitz (2011), Cherchye et al. (2012), Browning et al. (2013), Lise and Yamada (2014), and Cherchye et al. (2015) pointed out that intra-household bargaining power differs between husband and wife, and the intra-household bargaining power gap may influence household resource allocation. It is thought that the intra-household bargaining power gap and the relative income gap may affect married women's well-being, however published empirical studies on the issue are scarce. Most published studies analyze the determinants of well-being based on the Ordinary Least Squares (OLS) or ordered logit regression model using one period or repeated cross section survey data, and there may be an heterogeneity problem in these results.

This study uses an empirical study to proof the absolute income hypothesis and relative income hypothesis using data from the Japanese longitudinal survey conducted from 1994 to 2014. The influence of intra-household bargaining power on the happiness of Japanese married women is examined. These results may provide new evidence for the study of happiness, and lead to a deeper understanding of the work-life conflict for Japanese married women.

This study is structured as follows. Firstly, we summarize the economic theory about the influence of absolute income and relative income on happiness and previous empirical studies on the issue. Secondly, we describe the methods of analysis, including introduction to models and data. Then, we give the calculated results, and

---

[1]Two hypotheses are concerned with the influence of relative income on happiness. The interdependence preference hypothesis for which Leibenstein (1950), Kapteyn et al. (1978), Frank (1985) pointed out that because the good satisfaction of the consumer is not only related to the good function itself, but also with non-good function need (e.g. the rise of social position through holding a high quality or high-price good), the owned good gap between the individual and his (her) reference group with similar characteristics (e.g. age, education) could influence subjective happiness. The second is the relative deprivation hypothesis. Easterlin (1974), Boskin and Sheshinski (1978), Layard (1980), Frank (1985) and Akerlof and Yellen (1990) emphasized that when the gap between the individual and his (her) reference group is greater, for example, and the income of the individual is lower than his (her) reference group, the individual might feel the inferior, which might cause unhappiness.

state and interpret the econometric results. Lastly, we present the main conclusions and policy implication.

1 Literature Review and Hypotheses

The relation between income and happiness, absolute income hypothesis and relative income hypothesis has been the subject of extensive published research. Firstly, the absolute income hypothesis finds that subjective happiness is higher for the high-income group than for the low-income group. The absolute hypothesis is supported in most previous studies for both developed countries including Japan (Hamermesh 1977; Easterlin 2001; Ferrer-i-Carbonell 2005; Vendrik and Woltjer 2007; Sano and Otake 2007; Otake et al. 2010; Tsutsui 2010) and developing countries (Appleton and Song 2008; Smyth et al. 2010; Jiang et al. 2011; Wang and VanderWeele 2011; Ma 2016). However, it is also rejected for some developing countries including China (Luo 2006, 2009). The relative income hypothesis emphasizes how the gap between the individual (or household) and the reference group may affect the individual's subjective well-being; the probability of unhappiness is greater for those whose income level is lower than that of the reference group. In previous studies, the reference group is usually defined as having characteristics similar to the analyzed unit (individual or household), and the gaps between individual (or household) income and the average income of the reference group is usually utilized as the index of relative income. The relative income hypothesis is supported for both developed countries (Hamermesh 1977; Ferrer-i-Carbonell 2005; Vendrik and Woltjer 2007; Tsutsui 2010) and developing countries (Luo 2006, 2009; Brockmann et al. 2009; Wang and VanderWeele 2011; Ma 2016).

Irokawa (1999) undertook an empirical study using data from the Japanese Panel Survey of Consumers (JPSC) from 1995 to 1997, and finds the total income of wife and husband positively affects life satisfaction, therefore the absolute income hypothesis is supported. Urakawa and Matsuura (2007) analyze the influence of relative income on happiness using data from the JPSC for 1994 to 2001, and indicate that the relative income hypothesis is supported. Sakamoto (2008) analyzes the effect of the wife's work status, intra-household resource allocation (time and consumption) on happiness using data from the JPSC for 1994 to 2004, and finds that greater household income positively affects happiness, which means the absolute income hypothesis is supported. He finds that other factors, such as education, age, children, or return to work also influence happiness. Higuchi and He (2011) proof the relative income hypothesis using data from the JPSC for 1993 to 2009, and indicate that in Japan the relative income hypothesis is supported and the marriage status, work status, children, living with parents and economic fluctuation also influence happiness. Higuchi and Hagiwara (2011) employ an empirical study using JPSC data. They find the wife's income and the husband's income affect happiness, and other factors (e.g. age, marriage, and husband housework hours) also influence happiness.

Even though the absolute income and relative income hypotheses are estimated in previous studies, some issues remain to be discussed. The concerns of this study are as follows: first, for the index of relative income, the gap between individual (or household) income and reference group which is usually used in previous studies which can be defined as "inter-household relative income", whereas another kind of relative income, for example, the income gap between wife and husband which is relation with intra-household bargaining power is not considered. In the study, we will utilize various index of intra-household bargaining power[2] to extend the content of the relative income hypothesis, and proof the relative income hypotheses from various perspectives, which are not considered in the previous studies. In this study, based on the survey items of the questionnaires, we utilize the income (or wage), education attainment of the wife and her husband, and the information about husband income controlled by the wife to conduct the index of intra-household relative income. Second, this study utilizes a panel data analysis model to address the heterogeneity problem. A dynamic panel data analysis method is employed to find out if an initial dependent problem exists (Wooldridge 2005). New findings based on these econometric analyses develop the previous studies.

Happiness economic theory and published research are used to develop an empirical study to test the three hypotheses.

[H1: Absolute Income Hypothesis]
The happiness of married women is greater for the higher income (wage) group than for the middle and low income (wage) groups in Japan.

[H2: Inter-household Relative Income Hypothesis]
The happiness of Japanese married women is greater for the group whose income (wage) is higher than the income of the reference group.

[H3: Intra-household Bargaining Power Hypothesis]
Even though the bargaining power gap between wife and husband may affects the happiness, there exits two effects. Concretely, the more bargaining power a wife has, the greater happiness the wife will feel (positive effect), whereas the inverse relationship may exist (negative effect). When the positive effect is greater for the negative effect, the bargaining power gap between wife and husband may positively affects the happiness of married women in Japan.

For H1 and H2, as described above, it is presumed that the absolute income level and the inter-household relative income will positively affect married women's happiness. However, it is not clear how the intra-household bargaining power gap affects married women's happiness. Two kinds of effects may influence the effect of intra-household bargaining power gap between wife and husband on the wife's happiness. First, the wife's intra-household bargaining power may increase when

[2]For the empirical studies on intra-household bargaining power, please refer Browning et al. (1994), Chiappori et al. (2002),Couprie (2007), Lise and Seitz (2011), Cherchye et al. (2012), Browning et al. (2013), Lise and Yamada (2014), Cherchye et al. (2015).

her income (or wage) increases. When the wife feels very happy with the increase of intra-household bargaining power, her happiness may become greater when her income (wage) is higher than her husband's (positive effect). Second, it is thought that working hours might be longer for the group with a high income (wage), particularly for the married women group. The high income (wage) married women group may work for a longer time, which might cause work-family conflict to decrease married women's happiness (negative effect). In addition, based on Confucianism and the marriage matching model, the patriarchal consciousness crystallized as "men for work, women for family" persists as an influence on Japanese women's marriage and labor participation behaviors. Thus when the income (wage) is higher for the wife than the husband, the wife may feel unhappiness (negative effect). When the positive effect is greater for the negative effect, the bargaining power gap between wife and husband may positively affects the happiness of married women in Japan. Thus, an empirical study is needed to proof if both positive effect and negative effects obtain for H3.

2 Data and Method

2.1 Data

This study uses the Japanese Panel Survey of Consumers (JPSC) data. JPSC was first conducted in 1993 by randomly selecting young women aged 24–34 years old as Cohort A. Cohort B was added in 1997 for women aged 24–27. In 2003 Cohort C was added for women aged 24–29. Lastly, Cohort D was added for women aged 24–28. The JPSC was conducted every year from 1993 to 2017. Because the survey objects are effectively controlled and managed, the samples for the panel survey provide long-term balance panel datasets for this study. The JPSC questionnaire includes information about the subjective happiness of the wife, the household member's (wife and husband)'s demographics (i.e. age, education) and household characteristics (i.e. number of children, children's age, hours of husband participation in child care or housework). Detailed information on income (i.e. wife and husband's yearly income and wage, household yearly income, work status, intra-household income transfer and household income management pattern) can be obtained. Using this information, the empirical study for H1, H2, H3 hypotheses tests can be employed. The samples utilized in the study are married couple (wife and husband) for 19 waves, which is from 1995 to 2013.[3] The observations with missing values are deleted.

[3]Even though the JPSC is conducted from 1993 to 2013, the information of married women happiness can obtained only for 19 waves, which is from 1995 to 2013, therefore the panel dataset from 1995 to 2013 is utilized in this study.

2.2 Models

Fixed effects and random effects models are used in this study to investigate the relation between income, intra-household bargaining power and Japanese married women's happiness. In previous studies, when the dependent variable is an ordered category variable, the ordered logit regression model is usually utilized; when the dependent variable is a continuous variable, the OLS and the panel data analysis methods are used. The estimated results based on these methods are consistent. Because when the dependent variable is a continuous variable the results are easily understood, a continuous variable of the wife's subjective happiness score (very happy = 5, happy = 4, normal = 3, unhappy = 2, very unhappy = 1) is used as the dependent variable. The models are expressed as Eqs. (1) and (2).

$$H_i = a + \beta_1 \ln y_i + \beta_2 \ln(y_i/y_i^*) + \beta_X X_i + \varepsilon_i \tag{1}$$

$$H_{it} = a + \beta_1 \ln y_{it} + \beta_2 \ln(y_{it}/y_{it}^*) + \beta_X X_{it} + v_i + \varepsilon_{it} \tag{2}$$

Equation (1) represents the pooling OLS model. Equation (2) represents the fixed-effect model or random-effect model. In Eqs. (1) and (2), i denotes wife individual, t denotes survey year (from 1995 to 2013), H is the wife happiness score from 1 to 5, $\ln y$ is the logarithmic value of household (or wife) income variable, $\ln y/y^*$ are a set of logarithmic value of relative income including inter-household relative income and intra-household relative income. a is a constant, β are the estimated coefficients. ε is a true error. v is an item related with individual specific and time invariant factors. In Eq. (1), because v_i is not considered, heterogeneity problems may occur in the estimated results. In Eq. (2), because the fixed effects model and random effects model is based on first difference (FD) estimation, v_i will drop out by first difference (FD) estimation, thus the heterogeneity problem can be addressed by the fixed effects model and random effects model.

H1, H2 and H3 are proofed using the results of β_1 and β_2. When β_1 is positive and it is statistically significant, H1 is supported. When $\ln y/y^*$ is inter-household relative income index, β_2 is positive value, and it is statistically significant, H2 is supported. When $\ln y/y^*$ is intra-household relative income index, β_2 is statistically significant, H3 is supported.

As pointed out in Wooldridge (2002, 2005) and Contoyannis et al. (2004), there may be an initial dependent problem in Eq. (2). The happiness in $t - 1$ period might affect happiness in t period. To address the problem, the dynamic fixed-effect (or random-effect) model is used in this study. It is expressed by Eq. (3).

$$H_{it} = a + \beta_{Ht-1} H_{i(t-1)} + \beta_1 \ln y_{it} + \beta_2 \ln(y_{it}/y_{it}^*) + \beta_X X_{it} + v_i + \varepsilon_{it} \tag{3}$$

In Eq. (3), H_{t-1} denotes happiness in the $t - 1$ period. The definitions of the others are similar for Eq. (2).

These results based on Eqs. (1)–(3) are utilized to proof hypotheses H1, H2 and H3.

The F test, the Breusch and Pagan Lagrangian multiplier test, and the Hausman specification test are employed in order to compare which model is available among the pooling OLS model, fixed effects model and random effects model. Particularly, the Breusch-Pagan Lagrangian multiplier test should be to test whether the variance of individual (unique) effects is zero. The Hausman test is used to test whether the correlation between individual effects and explanatory variables are zero.

2.3 Variable Setting

A continuous variable uses the wife subjective happiness score (SHS) for dependent variable setting. It is calculated as "very happy = 5, happy = 4, normal = 3, unhappy = 2 and very unhappy = 1".[4]

The important independent variables are the index for H1, H2, H3 hypotheses. They are constructed as follows, see Table 1:

First, household income is utilized as the index of absolute income to proof H1. 1st to 5th income quintiles dummy variables and wife wage are constructed. Wife wage is utilized to make the robustness check for H1.

Second, (1) household income gap (the ratio of the wife's household income to the reference group household income), is used as the inter-household relative income index for H2. (2) Wife wage gap (the ratio of wife wage to reference group wife wage) is used to make the robustness check for H2. The household income and wife wage of the reference group are an imputed value calculated from the income and wage functions.[5]

Third, in published studies based on the collective model the income index of intra-household bargaining power is utilized as follows: (1) wage gap, which is the ratio of wife hourly wage to husband hourly wage (or to the total hourly wage of wife and husband) (Chiappori et al. 2002; Couprie 2007; Cherchye et al. 2012, 2015); (2) income gap, which is the ratio of wife income to husband income (or to the total income of wife and husband (Browning et al. 1994, 2013); (3) non-earning income gap which is the ratio of wife's non-earning income to husband's non-earning income (or to the total non-earning income of wife and husband) (Chiappori et al. 2002; Couprie 2007; Cherchye et al. 2012, 2015); and (4) saving gap which is the

[4]Value is transformed into opposite order based on the questionnaire item. Although the value for very happy is "1" in the survey questionnaire, it is transformed to "5" in the present analysis.

[5]For household income function, the dependent variable is household equivalent income, independent variables are husband's experience year, education attainment, work status, cities and countries scale. For wife wage function, the dependent variable is wife wage, independent variables are wife's experience year, education attainment, work status, cities and countries scale.

Table 1 Index of H1, H2, H3 hypotheses

Hypotheses	Index	Contents
H1: Absolute Income Hypothesis	i. Household income	1st to 5th income quintiles dummy variables
	ii. Wife wage	Wife wage
H2: Inter-household Relative Income Hypothesis	i. Household income gap	Ratio of wife's household income to reference group household income (imputed value)
	ii. Wife wage gap	Ratio of wife wage to reference group wife wage (imputed value)
H3: Intra-household Relative Income Hypothesis	i. Income gap	Ratio of wife income to total income of wife and husband
	ii. Controlled income	Proportion of the controlled husband income by wife to total husband income (or couple income)
	iii. Wage gap	Ratio of wife wage to total wage of wife and husband
	iv. Education gap	Gap of wife education level to husband education level

Note
1. Household income is utilized for H1 test
2. Household income gap is utilized for H2 test; income gap and controlled income by wife are utilized for H3 test
3. The other variables are utilized n robustness check estimations for these hypotheses
Source By authors

ratio of wife's saving to husband's saving (or to the total saving of wife and husband) (Lise and Yamada 2014).[6]

To refer the previous studies and to utilize the JPSC questionnaires, the fourth type of variables are utilized as the indices of the intra-household bargaining power to proof H3. They are as follows: (1) the intra-household income gap, which is the ratio of wife income to total household income (the total income of wife and husband); (2) the ratio of controlled husband income by wife, which is the proportion of controlled household income by wife to total husband disposable income; (3) the intra-household wage gap, which is the ratio of wife wage to total household wage (total wage of wife and husband)[7]; and (4) the intra-household education gap,[8]

[6]The age and education gaps between wife and husband are also utilized for the index of intra-household bargaining power in the previous studies (e.g. Browning et al. 1994; Lise and Yamada 2014).

[7]It is equal to "0" when the individual is non-work.

[8]The education attainment level is evaluated as follows: junior high school graduation is equal to 1; vocational school graduation (entrance requirement junior high school graduation) is equal to 2;

which is the education gap between wife and husband. Particularly, (2) the controlled income rate is an original index utilized for the first issue. It is conducted as follows: the questionnaire asks the wife whether her husband transferred all of his disposable income to her (yes or no) and how much is the transferred income. The transferred husband income account is used as the husband's income account controlled by the wife. It is thought when the intra-household bargaining power is large for the wife, the proportion of husband income controlled by the wife will be large.

The household equivalent income utilized in the study is calculated based on an equivalent coefficient.[9] In addition, the income (household, wife, and husband) from 1995 to 2013 is adjusted by the Japanese consumption price index (CPI) from 1995 to 2013 which are published by the Ministry of Internal Affairs and Communications, Japan. The CPI in 1995 provides the standard.

Fourth, the other variables (controlled variables) are also conducted, they are (1) wife's age dummy variables, which are the groups aged 24–29, 30–39 and older than 40 years[10]; (2) wife education dummy variables, which are the highest education attainment: they are the junior high school, senior high school, vocational school, college and university or the graduate school of the university; (3) wife work status dummy variables, which include the non-work, irregular worker, regular worker, the self-employed and others; (4) youngest child's age and children number; (5) weekly husband child care or housework hours[11]; (6) living with parents (either wife's parents or husband's parents) dummy variable; (7) home status (rent room or my home) dummy variable; (8) city and country scale dummy variables, which is divided into the household lives in large city, the other city or countryside; (9) survey year dummy variables, the 2008 dummy variable is used to control the influence of the world financial crisis.

The statistical description of dependent variable and independent variables are summarized in Table 2.

3 Situations of Absolute Income/Relative Income and the Happiness of Married Women in Japan

First, Table 3 summarizes the happiness score distribution of Japanese married women by household income categories. Generally, in the long-term (the period from 1995 to 2013), although the proportion of the very happy group is highest for the highest income group (20.7% for the income fifth quintile group), the differences in the proportions between income categories (from income first quintile to income

---

senior high school graduation is equal to 3; vocational school graduation is (entrance requirement senior high school graduation) is equal to 4; college school graduation is equal to 5; university graduation is equal to 6, and graduate school graduation is equal to 7.

[9]In the study, the square root of family numbers is utilized as the equivalent coefficient.

[10]In this dataset, the youngest age of a wife is 24 years old.

[11]It is calculated by assuming 5 ordinary days and 2 days holiday per week.

Table 2 Descriptive statistics

Variable type		Total	Happy	Normal	Unhappy
Indices for H1	Household income	368	374	358	324
	Wife wage	107	106	110	111
Indices for H2	Household income gap	0.986	0.996	0.971	0.893
	Wife wage gap	0.641	0.625	0.679	0.733
Indices for H3	Income gap (wife/husband)	0.152	0.146	0.168	0.180
	Proportion of controlled husband income by wife to husband income	0.853	0.855	0.852	0.836
	Proportion of controlled husband income by wife to couple income	0.695	0.701	0.681	0.660
	Wage gap (wife/husband)	0.150	0.144	0.167	0.180
	Education gap (wife–husband)	−0.164	−0.225	0.016	0.125
Control variables	*Education*				
	Junior high school	0.041	0.035	0.055	0.089
	Senior high school	0.407	0.390	0.468	0.459
	Vocational school	0.184	0.183	0.193	0.174
	College	0.227	0.233	0.199	0.217
	University	0.141	0.159	0.085	0.061
	Age				
	Aged 24–29 years	0.127	0.142	0.070	0.089
	Aged 30–29 years	0.522	0.537	0.474	0.450
	Aged more than 40 years	0.352	0.321	0.456	0.461
	Number of children				
	No child	0.099	0.104	0.090	0.066
	One	0.222	0.231	0.198	0.181
	Two	0.480	0.474	0.493	0.517
	More than three	0.199	0.192	0.218	0.236
	Youngest child age				
	No child	0.099	0.104	0.090	0.066
	Aged 0–3 years	0.300	0.333	0.192	0.178
	Aged 4–6 years	0.160	0.160	0.158	0.169
	Aged 7–14 years	0.298	0.278	0.359	0.390
	Aged more than 15 years	0.142	0.126	0.201	0.196

(continued)

Table 2 (continued)

Variable type		Total	Happy	Normal	Unhappy
	Hours of husband participation in child care or homework	8.624	9.183	6.944	6.029
	Work status				
	Non-work	0.444	0.467	0.368	0.355
	Regular worker	0.200	0.204	0.192	0.182
	Irregular worker	0.300	0.280	0.360	0.391
	Self-employed and other	0.056	0.050	0.080	0.072
	Living with parents	0.282	0.274	0.306	0.317
	House status				
	Rent	0.302	0.304	0.291	0.308
	Home owner	0.698	0.696	0.709	0.692
	Cities and countries scale				
	Large city	0.256	0.255	0.245	0.300
	Middle-small city	0.619	0.621	0.624	0.573
	Countryside	0.125	0.124	0.131	0.127
Observations		14,780	11,446	2,460	874

Note
1. Mean values are summarized in Table 2
2. For the definition of index for H1–H3, please see Table 1
3. The observations for account of wife' controlled income to couple income is 12,592 because a part of wives' September incomes are missing values
Source Calculated based on JHPS 1995–2013

fifth quintile) are small. To compare the situation for the three periods 1995–1999, 2000–2007, and 2008–2013, the proportions of the very happy group are higher for the income fifth quintile group than for the other income quintile groups (from income first quintile group to income fourth quintile group) for both the 1995–1999 and 2000–2007 periods; whereas the differences of the proportions of the very happy group between each income category are small for the 2008–2013 period. It is indicated that in the period before the world financial crisis, the probability of being very happy may be higher for the high-income group than for the middle and low income groups, but the influence of absolute income on Japanese married women's happiness decreased after the world financial crisis. It indicates the effect of absolute income on Japanese married women's happiness might greater for the period before the world financial crisis, but the effect is less in the present.

Table 4 summarizes the happiness score distribution of Japanese married women by inter-household relative income category. Here, I_0 gives a reference value for average household income of the group with similar characteristics which is calculated based on the household income function (see footnote5). I_1 is the actual value of household income. $I_1 \geq I_0$ expresses the wife's household income is higher than

Table 3 Happiness score distribution of Japanese married women by household income categories

	Very happy	Happy	Normal	Unhappy	Very unhappy	% Total
1995–1999						
Income first quintile	20.0	60.6	14.1	3.7	1.5	100.0
Income second quintile	23.9	60.5	11.3	3.8	0.5	100.0
Income third quintile	24.2	56.6	15.2	3.5	0.5	100.0
Income fourth quintile	22.8	64.5	9.3	3.3	0.3	100.0
Income fifth quintile	25.5	62.5	10.8	1.0	0.3	100.0
2000–2007						
Income first quintile	16.9	52.9	20.9	6.9	2.5	100.0
Income second quintile	19.7	55.1	18.9	5.0	1.3	100.0
Income third quintile	18.1	55.9	18.6	6.0	1.5	100.0
Income fourth quintile	19.7	59.1	16.7	3.6	1.0	100.0
Income fifth quintile	20.6	62.0	12.9	3.9	0.7	100.0
2008–2013						
Income first quintile	21.6	50.5	18.0	7.1	2.8	100.0
Income second quintile	20.1	56.4	17.1	5.7	0.7	100.0
Income third quintile	18.6	57.9	18.0	4.5	1.0	100.0
Income fourth quintile	20.9	58.6	16.0	3.7	0.9	100.0
Income fifth quintile	20.0	59.7	16.4	3.1	0.8	100.0
1995–2013						
Income first quintile	19.5	53.1	18.6	6.4	2.4	100.0
Income second quintile	20.6	55.8	17.1	5.2	1.2	100.0
Income third quintile	19.0	57.2	17.8	5.0	1.0	100.0

(continued)

Table 3 (continued)

	Very happy	Happy	Normal	Unhappy	Very unhappy	% Total
Income fourth quintile	20.5	59.5	15.5	3.6	0.8	100.0
Income fifth quintile	20.7	61.1	14.2	3.2	0.7	100.0

Note
1. Household equivalent income is used. The equivalence index is square root of household numbers
2. The observations are 2,002, 6,705, and 6,073 for the periods during the 1995–1999, the 2000–2007, and the 2008–2013, respectively
Source Calculated based on JHPS 1995–2013

Table 4 Happiness score distribution of Japanese married women by inter-household relative income categories

	Very happy	Happy	Normal	Unhappy	Very unhappy	% Total
1995–1999						
I1 ≥ I0	24.1	59.9	12.9	2.8	0.4	100.0
I1 < I0	22.8	61.5	11.7	3.2	0.7	100.0
2000–2007						
I1 ≥ I0	20.9	58.0	15.9	4.2	1.1	100.0
I1 < I0	17.8	56.4	18.6	5.6	1.6	100.0
2008–2013						
I1 ≥ I0	22.5	57.7	15.6	3.4	0.8	100.0
I1 < I0	18.7	55.9	18.1	5.7	1.6	100.0
1995–2013						
I1 ≥ I0	21.9	58.1	15.4	3.7	0.9	100.0
I1 < I0	18.9	56.9	17.4	5.3	1.5	100.0

Note
1. I1: household equivalent income of married women
2. I0: average household income of the group with the similar characteristics, which is an imputed value based on household income function
Source Calculated based on JHPS 1995–2013

the household income of the reference group. Both in the long-term (1995–2013) and in the short-term (1995–1999, 2000–2007, and 2008–2013), the proportions of the very happy group are higher for the high inter-household relative income group ($I_1 \geq I_0$ group), they are 21.9% for 1995–2013, 24.1% for 1995–1999, 20.9% for 2000–2007, and 22.5% for 2008–2013. It indicates that inter-household relative income may affect the happiness of married women in Japan.

Finally, to consider the relation between intra-household relative income and happiness, the happiness score distribution of Japanese married women is calculated by intra-household relative income categories, these results are summarized in Table 5. Two types of intra-household relative income index are used. The first compares the wife income and husband income, the second is the proportion of husband income controlled by the wife. When the income is higher for the wife than the husband (W > H group), and when the proportion of controlled husband income is higher (from 60% to less than 80% group, more than 80% group), the proportion of very happy is smaller than for the other groups. However, when the wife's income is equal to her husband's income, the proportion of very happy is highest in each period.

The state of income and happiness for Japanese married women based on these calculated statistics results may be observed but other factors that affect happiness (e.g. wife's characteristic, family structure) are not controlled. H1, H2, H3 hypotheses are tested using econometric analysis methods in the following section.

4 Econometric Analysis Results: Three Hypotheses Tests

4.1 Results of Hypothesis Tests

The results of three hypotheses tests are shown in Tables 6, 7, and 8. (1) The Pooling OLS, (2) the fixed effects model, and (3) the random effects model are used in the analysis. The results based on the F-test and the Breusch and Pagan Lagrangian multiplier test indicate that both the fixed effects model and the random effects model are more propriety than the OLS; the results based on Hausman specification test indicate that the fixed effects model is more propriety than the random effects model. Therefore we will discuss the hypotheses test using the results based on the fixed effects model in Tables 6, 7 and 8. The main findings are as follows.

First, for H1, the results show that for the low income group (income first quintile), the happiness score is higher than for high income group (income fourth and fifth quintile groups). For example, based on the results in Table 7, the coefficients of both fourth quintile and fifth quintile income groups are positive values (0.070 and 0.111), and they are statistically significant. These results indicate that the happiness of Japanese married women is greater for the higher income group, therefore the absolute income hypothesis is supported. The result is consistent with the previous studies for other developed countries (Hamermesh 1977; Ferrer-i-Carbonell 2005; Vendrik and Woltjer 2007; Sano and Otake 2007; Otake et al. 2010; Tsutsui 2010) and for developing countries (Appleton and Song 2008; Smyth et al. 2010; Jiang et al. 2011; Wang and VanderWeele 2011; Ma 2016).

Second, for H2, the coefficient of $I_1 \geq I_0$ dummy variable is not statistically significant (see Table 6). H2 is tested by using the other two indexes of inter-household relative income: the ratio of I_1 to I_0 (I_1/I_0), and the income gap account between I_1 and I_0 (gap $= I_1 - I_0$), these results are similar with those in Table 6. Based on these results, the H2 [the happiness of Japanese married women is greater for the group whose income (wage) is higher than the reference group] is not supported when

Table 5 Happiness score distribution of Japanese married women by intra-household relative income categories

	Very happy	Happy	Normal	Unhappy	Very unhappy	% Total
A: Income gap						
1995–1999						
W > H	15.4	65.4	7.7	11.5	0.0	100.0
W = H	30.0	55.0	10.0	5.0	0.0	100.0
W < H	23.3	60.9	12.2	2.9	0.6	100.0
2000–2007						
W > H	19.2	56.5	16.1	6.2	2.1	100.0
W = H	24.3	48.6	16.2	10.8	0.0	100.0
W < H	18.9	57.1	17.6	5.0	1.4	100.0
2008–2013						
W > H	23.8	52.0	19.0	4.8	0.4	100.0
W = H	34.9	39.5	20.9	2.3	2.3	100.0
W < H	20.0	56.9	17.0	4.8	1.3	100.0
1995–2013						
W > H	21.4	54.6	17.2	5.7	1.1	100.0
W = H	30.0	46.0	17.0	6.0	1.0	100.0
W < H	20.0	57.5	16.6	4.6	1.2	100.0
B: Controlled income						
1995–1999						
Less than 20%	24.2	54.5	14.1	6.1	1.0	100.0
20% to less than 40%	20.9	61.5	15.4	2.2	0.0	100.0
40% to less than 60%	22.6	68.7	8.7	0.0	0.0	100.0
60% to less than 80%	18.5	71.6	7.4	1.2	1.2	100.0
More than 80%	23.6	60.2	12.3	3.2	0.6	100.0
2000–2007						
Less than 20%	19.6	53.1	16.6	7.1	3.6	100.0
20% to less than 40%	22.9	56.0	15.6	4.5	1.0	100.0
40% to less than 60%	17.1	59.0	18.1	4.5	1.3	100.0
60% to less than 80%	18.4	57.5	20.3	3.8	0.0	100.0
More than 80%	18.8	57.2	17.6	5.0	1.3	100.0

(continued)

Table 5 (continued)

	Very happy	Happy	Normal	Unhappy	Very unhappy	% Total
2008–2013						
Less than 20%	23.4	50.4	18.1	5.7	2.3	100.0
20% to less than 40%	26.1	54.6	14.7	3.5	1.1	100.0
40% to less than 60%	16.6	57.1	20.6	5.1	0.7	100.0
60% to less than 80%	14.3	62.2	17.4	4.8	1.3	100.0
More than 80%	19.9	57.1	17.0	4.8	1.2	100.0
1995–2013						
Less than 20%	22.0	51.9	17.1	6.3	2.7	100.0
20% to less than 40%	24.2	55.9	15.1	3.8	1.0	100.0
40% to less than 60%	17.8	59.8	17.6	4.0	0.8	100.0
60% to less than 80%	16.8	61.4	17.3	3.8	0.7	100.0
More than 80%	19.9	57.6	16.6	4.7	1.2	100.0

Note W: wife income; H: husband income
Source Calculated based on JHPS 1995–2013

controlled by the other factors including the absolute income. The results may be due to the influence of absolute income which is greater than inter-household relative income. This will be discussed in more detail using the robustness check results.

Third, for H3, (1) based on the results in Table 7, the coefficient of income gap between wife and husband is a negative value (−0.198), and it is statistically significant at 1%. It indicates that when the income gap between wife and husband increases, the wife's happiness may decrease. Thus H3 is not supported. It may be caused by the traditional gender responsibility consciousness. In Japan the traditional consciousness of the gender role "Men for work, Women for Family" persists. Thus the husband will gain more income in the job market, whereas the wife's main responsibility is to do housework or care for children and other family members (e.g. patient parent care). When income in the job market is higher for the wife than for the husband but the wife's housework or family care work responsibility is still greater than the husband's, it can be expected that the work-family conflict may be greater for the married women with a high wage.[12] When the wife's income is higher than the husband's income, the wife may experience unhappiness. (2) Based

[12]Ma (2007) indicated that housework and family care time for the non-work wife and working wife is almost the same in Japan. She points out that compared with husbands in other developed countries Japanese husbands do little housework.

Table 6 Results of absolute income hypothesis (H1) and inter-household relative income hypothesis (H2) tests

	(1) Pooling		(2) Fixed effects		(3) Random effects	
	Coeff.	S.E.	Coeff.	S.E.	Coeff.	S.E.
Household income (ref. income first quintile)						
Income second quintile	0.044**	0.018	0.009	0.021	0.036*	0.019
Income third quintile	0.033*	0.019	0.016	0.025	0.036*	0.021
Income fourth quintile	0.098***	0.022	0.060*	0.031	0.105***	0.025
Income fifth quintile	0.140***	0.028	0.101**	0.040	0.146***	0.031
Household income gap (ref. $I_1 < I_0$)						
$I_1 > I_0$	−0.026	0.018	−0.004	0.022	−0.021	0.020
t − 1Happiness (ref. very unhappy)						
Very happy	0.025***	0.001	0.007***	0.001	0.020***	0.001
Happy	0.019***	0.001	0.006***	0.001	0.015***	0.001
Normal	0.012***	0.001	0.004***	0.001	0.009***	0.001
Unhappy	0.007***	0.001	0.002***	0.001	0.005***	0.001
Education (ref. junior high school)						
Senior high school	0.081***	0.030			0.126***	0.040
Vocational school	0.093***	0.032			0.148***	0.042
College	0.104***	0.032			0.165***	0.042
University or more	0.150***	0.034			0.220***	0.044
Age (ref. aged 24–29 years)						
Aged 30–39 years	−0.037*	0.022	−0.091***	0.028	−0.065***	0.023
Aged 40 years and older	−0.074***	0.027	−0.088**	0.038	−0.108***	0.029
Number of children (ref. no child)						
One	−0.010	0.028	0.020	0.048	−0.001	0.031
Two	−0.020	0.027	−0.025	0.052	−0.023	0.031
More than three	0.013	0.029	−0.002	0.064	0.009	0.034
Youngest child age (ref. aged 0–3 years and no-child)						
Aged 4–6 years	−0.018	0.018	−0.034*	0.020	−0.030*	0.018
Aged 7–14 years	−0.022	0.019	−0.032	0.027	−0.047**	0.020

(continued)

Table 6 (continued)

	(1) Pooling		(2) Fixed effects		(3) Random effects	
	Coeff.	S.E.	Coeff.	S.E.	Coeff.	S.E.
Aged 15 years and older	−0.031	0.025	0.014	0.042	−0.065	0.028
Hours of husband participation in child care or homework	0.002***	0.001	0.002**	0.001	0.002***	0.001
Work status (ref. non-work)						
Regular worker	−0.043***	0.017	−0.052*	0.029	−0.045**	0.020
Irregular worker	−0.040***	0.014	−0.024	0.019	−0.045***	0.016
Self-employed and other	−0.070***	0.026	−0.074**	0.033	−0.081***	0.028
Living with parents	−0.039***	0.015	−0.083**	0.034	−0.051***	0.018
House status (ref. rent)						
Home owner	0.026*	0.014	0.037	0.026	0.027	0.017
Cities and countries scale	Yes		Yes		Yes	
Survey year	Yes		Yes		Yes	
Number of observations	10,954		10,954		10,954	
Number of groups	1,717		1,717		1,717	
R-sq. within			0.07		0.05	
Between			0.48		0.73	
Overall			0.31		0.45	
F-test that all $u_i = 0$			2.99 ($p = 0.0000$)			
Breusch and Pagan Lagrangian multiplier test for random effects					132.56 ($p = 0.0000$)	
Hausman specification test			3218.60 ($p = 0.0000$)			

Note *, **, ***: statistical significant levels are 10%, 5%, 1%
Source Calculated based on JHPS 1995–2013

Table 7 Results of absolute income hypothesis (H1) and intra-household relative income hypothesis (H3) tests (1)

	(1) Pooling		(2) Fixed effects		(3) Random effects	
	Coeff.	S.E.	Coeff.	S.E.	Coeff.	S.E.
Household income (ref. income first quintile)						
Income second quintile	0.043**	0.018	0.013	0.021	0.036*	0.019
Income third quintile	0.026	0.019	0.023	0.025	0.032	0.020
Income fourth quintile	0.082***	0.019	0.070**	0.028	0.094***	0.022
Income fifth quintile	0.113***	0.021	0.111***	0.034	0.129***	0.024
Income gap between wife and husband	−0.171***	0.055	−0.198***	0.075	−0.194***	0.059
t − 1Happiness (ref. very unhappy)						
Very happy	0.025***	0.001	0.007***	0.001	0.020***	0.001
Happy	0.019***	0.001	0.006***	0.001	0.015***	0.001
Normal	0.012***	0.001	0.004***	0.001	0.009***	0.001
Unhappy	0.007***	0.001	0.002***	0.001	0.005***	0.001
Education (ref. junior high school)						
Senior high school	0.087***	0.030			0.133***	0.040
Vocational school	0.101***	0.032			0.157***	0.042
College	0.113***	0.031			0.174***	0.042
University or more	0.160***	0.034			0.230***	0.044
Age (ref. aged 24–29 years)						
Aged 30–39 years	−0.036	0.022	−0.093***	0.028	−0.065***	0.023
Aged 40 years and older	−0.071***	0.027	−0.089**	0.038	−0.107***	0.029
Number of children (ref. no child)						
One	−0.020	0.028	0.004	0.048	−0.014	0.031
Two	−0.032	0.027	−0.043	0.053	−0.038	0.031
More than three	0.001	0.029	−0.023	0.065	−0.007	0.034
Youngest child age (ref. aged 0–3 years and no-child)						
Aged 4–6 years	−0.016	0.018	−0.030	0.020	−0.028	0.018
Aged 7–14 years	−0.018	0.019	−0.027	0.027	−0.043**	0.020

(continued)

Table 7 (continued)

	(1) Pooling		(2) Fixed effects		(3) Random effects	
	Coeff.	S.E.	Coeff.	S.E.	Coeff.	S.E.
Aged 15 years and older	−0.024	0.025	0.021	0.042	−0.057**	0.028
Hours of husband participation in child care or homework	0.002***	0.001	0.002**	0.001	0.002***	0.001
Work status (ref. non-work)						
Regular worker	0.012	0.025	−0.022	0.031	0.010	0.026
Irregular worker	−0.018	0.016	−0.011	0.020	−0.022	0.017
Self-employed and other	−0.038	0.028	−0.053	0.034	−0.047	0.030
Living with parents	−0.036**	0.015	−0.086**	0.034	−0.048***	0.018
House status (ref. rent)						
Home owner	0.023	0.014	0.035	0.026	0.024	0.017
Cities and countries scale	Yes		Yes		Yes	
Survey year	Yes		Yes		Yes	
Number of observations	10,954		10,954		10,954	
Number of groups			1,717		1,717	
R-sq. within			0.07		0.05	
Between			0.46		0.73	
Overall			0.30		0.45	
F-test that all $u_i = 0$			3.00 ($p = 0.0000$)			
Breusch and Pagan Lagrangian multiplier test for random effects					134.22 ($p = 0.0000$)	
Hausman specification test			3199.92 ($p = 0.0000$)			

Note *, **, ***: statistical significant levels are 10%, 5%, 1%
Source Calculated based on JHPS 1995–2013

Table 8 Results of absolute income hypothesis (H1) and intra-household relative income hypothesis (H3) tests (2)

	(1) Pooling		(3) Fixed effects		(3) Random effects	
	Coeff.	S.E.	Coeff.	S.E.	Coeff.	S.E.
Household income (ref. income first quintile)						
Income second quintile	0.043**	0.018	0.009	0.021	0.035*	0.019
Income third quintile	0.027	0.019	0.016	0.025	0.031	0.020
Income fourth quintile	0.085***	0.019	0.059**	0.028	0.094***	0.022
Income fifth quintile	0.118***	0.021	0.098***	0.033	0.130***	0.024
Proportion of controlled husband income by wife to couple income	0.040**	0.020	0.071**	0.033	0.055**	0.024
t − 1Happiness (ref. very unhappy)						
Very happy	0.025***	0.001	0.007***	0.001	0.020***	0.001
Happy	0.019***	0.001	0.006***	0.001	0.015***	0.001
Normal	0.012***	0.001	0.004***	0.001	0.009***	0.001
Unhappy	0.007***	0.001	0.002***	0.001	0.005***	0.001
Education (ref. junior high school)						
Senior high school	0.083***	0.030			0.128***	0.040
Vocational school	0.096***	0.032			0.151***	0.042
College	0.109***	0.031			0.170***	0.042
University or more	0.156***	0.034			0.226***	0.044
Age (ref. aged 24–29 years)						
Aged 30–39 years	−0.034	0.022	−0.091***	0.028	−0.062***	0.023
Aged 40 years and older	−0.068**	0.027	−0.086**	0.038	−0.103***	0.029
Number of children (ref. no child)						
One	−0.011	0.028	0.019	0.048	−0.003	0.031
Two	−0.022	0.027	−0.026	0.052	−0.026	0.031
More than three	0.012	0.029	−0.005	0.064	0.007	0.033
Youngest child age (ref. aged 0–3 years and no−child)						
Aged 4–6 years	−0.018	0.018	−0.034*	0.020	−0.031*	0.018

(continued)

Table 8 (continued)

	(1) Pooling		(3) Fixed effects		(3) Random effects	
	Coeff.	S.E.	Coeff.	S.E.	Coeff.	S.E.
Aged 7–14 years	−0.022	0.019	−0.032	0.026	−0.048**	0.020
Aged 15 years and older	−0.031	0.025	0.013	0.042	−0.066**	0.028
Hours of husband participation in child care or homework	0.002***	0.001	0.002***	0.001	0.002***	0.001
Work status (ref. non-work)						
Regular worker	−0.043***	0.017	−0.049*	0.029	−0.044**	0.020
Irregular worker	−0.041***	0.014	−0.025	0.019	−0.045***	0.016
Self-employed and other	−0.071***	0.026	−0.073**	0.033	−0.081***	0.028
Living with parents	−0.038**	0.015	−0.085**	0.034	−0.049***	0.018
House status (ref. rent)						
Home owner	0.024*	0.014	0.037	0.026	0.026	0.017
Cities and countries scale	Yes		Yes		Yes	
Survey year	Yes		Yes		Yes	
Number of observations	10,954		10,954		10,954	
Number of groups	1,717		1,717		1,717	
R-sq. within			0.07		0.05	
Between			0.47		0.73	
Overall			0.31		0.45	
F-test that all $u_i = 0$			3.00 ($p = 0.0000$)			
Breusch and Pagan Lagrangian multiplier test for random effects					134.31 ($p = 0.0000$)	
Hausman specification test			3222.49 ($p = 0.0000$)			

Note *, **, ***: statistical significant levels are 10%, 5%, 1%
Source Calculated based on JHPS 1995–2013

on the results in Table 8, the coefficient of the proportion of controlled husband income is a positive value (0.071), and it is statistically significant at 1%. It is clear that when the proportion of controlled husband income increases, the wife's happiness increases. These results indicate that both intra-household bargaining power and traditional gender responsibility consciousness could influence the happiness of Japanese married women. Therefore H3 is supported. The results shown in Tables 7 and 8 indicate that H3 is partly supported as the hypothesis test results differ by the various bargaining power indexes.

Other factors also influence the happiness of Japanese married women. (1) The results show that the prior period happiness status positively affects present happiness. It shows it is appropriate to use the dynamic panel data analysis method. (2) The results based on the random effects model show that the high education group (particular for university and graduate school) is more likely to be happy than the low education group. (3) The probability of experiencing happiness is lower for the groups aged older than 30 than for the youth group aged 24 to 29. (4) The probability of happiness is lower for the group with a child under 3 years than for the group with a child aged 0 to 3 months. (5) The wife's happiness increases when the husband's housework hours become longer. It indicates that a husband doing more housework may increase the wife happiness. (6) Compared with the non-work group, the probability of happiness is lower for the working wife (to be regular worker, irregular worker, the self-employed and other employment status). (7) Living with parents may decrease the wife's happiness.

4.2 Robustness Check of Hypotheses Test

It is thought that the various income indexes and models may influence the hypotheses testing results. A set of robustness checks are made to further test the H1, H2, H3 hypotheses. In these models, the controlled variables are similar: they are education, age dummy, number of children, youngest child's age, hours of husband participation in child care or housework, work status, living with parents, house status, cities and countries scale, and survey year dummy variables. The index of absolute income and relative income is altered in model 1 to model 16. The results are summarized in Table 9. Four major conclusions emerge.

First, the results in model 1 show the low-income group (the income first quintile group) happiness score is higher than for high-income group (the income fourth quintile and fifth quintile groups). The result is the similar to that in Table 6. H1 is again supported.

Second, the results in model 2 indicate that when absolute income is not controlled, the household income gap positively affects the happiness at a 10% statistical level. H2 is supported. The results are different from those shown in Table 6. It indicates that the influence of the inter-household relative income on happiness is smaller than that of absolute income. The results may be explained in two ways. Firstly, the deregulation of irregular work and economic recession since the 1990s led to the

Table 9 Summary of robustness check for H1, H2, H3 hypotheses tests

Hypotheses	Index	Coeff.	S.E.
Model 1 (H1)	Household income + Con.		
	Household income (ref. income first quintile)		
	Income second quintile	0.008	0.021
	Income third quintile	0.015	0.025
	Income fourth quintile	0.058**	0.028
	Income fifth quintile	0.097***	0.033
Model 2 (H2)	Household income gap + Con.	0.034*	0.020
Model 3 (H3)	Income gap between wife and husband + Con.	−0.158**	0.074
Model 4 (H3)	Proportion of controlled husband income by wife to total husband income + Con.	0.071**	0.033
Model 5 (H3) (non-working wife)	Proportion of controlled husband income by wife to total husband income + Con.	0.049	0.056
Model 6 (H3) (working wife)	Proportion of controlled husband income by wife to total husband income + Con.	0.101**	0.046
Model 7 (H3)	Proportion of controlled husband income by wife to couple income + Con.	0.109***	0.040
Model 8 (H3) (non-working wife)	Proportion of controlled husband income by wife to couple income + Con.	0.047	0.065
Model 9 (H3) (working wife)	Proportion of controlled husband income by wife to couple income + Con.	0.164***	0.057
Model 10 (H3)	Wage gap between wife and husband + Con.	−0.168**	0.072
Model 11 (H3)	Education gap between wife and husband + Con.	−0.035***	0.005
Model 12 (H1 + H2 + H3)	[Household income + Inter-household income gap + Income gap between wife and husband + Con.]		
	Household income (ref. income first quintile)		
	Income second quintile	0.014	0.022
	Income third quintile	0.025	0.026
	Income fourth quintile	0.073**	0.032

(continued)

Table 9 (continued)

Hypotheses	Index	Coeff.	S.E.
	Income fifth quintile	0.117***	0.042
	Inter-household income gap	−0.005	0.026
	Income gap between wife and husband	−0.197***	0.075
Model 13 (H1 + H2 + H3)	[Household income + Household income gap + Proportion of controlled husband income by wife to couple income + Con.]		
	Household income (ref. income first quintile)		
	Income second quintile	0.011	0.022
	Income third quintile	0.019	0.026
	Income fourth quintile	0.065**	0.032
	Income fifth quintile	0.108**	0.042
	Inter-household income gap	−0.010	0.026
	Proportion of controlled husband income by wife to couple income	0.071**	0.033
Model 14 (H1 + H2 + H3)	[Wife wage + inter-household wage gap + wage gap between wife and husband + Con.]		
	Wife wage	0.000***	0.000
	Inter-household wife gap	0.000	0.002
	Wage gap between wife and husband	−0.388***	0.108
Model 15 (H1 + H2 + H3)	[Household income + Inter-household income gap + education gap between wife and husband + Con.]		
	Household income (ref. income first quintile)		
	Income second quintile	0.030	0.020
	Income third quintile	0.020	0.022
	Income fourth quintile	0.076***	0.025
	Income fifth quintile	0.106***	0.034
	Inter-household income gap	−0.008	0.021
	Education gap between wife and husband	−0.029***	0.006

(continued)

Table 9 (continued)

Hypotheses	Index	Coeff.	S.E.
Model 16 (H1 + H2 + H3)	[Household income + Inter-household income gap + Proportion of controlled husband income by wife to couple income + Con.]		
	Household income (ref. income first quintile)		
	Income second quintile	−0.004	0.024
	Income third quintile	0.013	0.029
	Income fourth quintile	0.045	0.035
	Income fifth quintile	0.090**	0.045
	Inter-household income gap	−0.002	0.027
	Proportion of controlled husband income by wife to couple income	0.112***	0.040

Note
1. *, **, ***: statistical significant levels are 10%, 5%, 1%
2. Controlled variables are expressed as "Con." in the table. They are education, age dummy, number of children, youngest child age, hours of husband participation in child care or homework, work status, living with parents, house status, cities and countries scale and survey year dummy variables. Although controlled variables are estimated in each model, they are not expressed in the table
3. The random effect is utilized for Model 4 and Model 15 due to the education gap is time invariant. Fixed effect model is utilized for other models
Source Calculated based on JHPS 1995–2013

increase of irregular worker, economic recession, the working poor group appeared and the relative poverty problem become serious, therefore the effect of absolute income on the happiness of married women is greater. Secondly, it may be caused by that inter-household income inequality is smaller in Japan.

Third, model 3 to model 11 used various indexes of intra-household relative income and various subsamples. Specifically, (1) when absolute income is not controlled the income gap (model 3), wage gap (model 10) and education gap (model 11) between wife and husband negatively affect the wife's happiness, whereas the proportion of controlled husband income positively influence the wife's happiness (model 4, model 7). These results are statistically significant at the 1% and 5% levels. These results are consistent with those in Table 8. Generally, when the proportion of controlled husband income to wife's income increases Japanese married women may feel increased happiness, whereas for the married women with high ability or high potential ability to earn more income (e.g. high income group, high education level group), work-family conflict may decrease happiness. H3 is supported again.

The total sample is divided into non-work wife group and working wife group in order to investigate in detail the influence of the proportion of controlled husband income. The results based on model 5, model 6, model 8, and model 9 indicate that the proportion of controlled husband income positively affects happiness significantly for the working wife group; whereas the results for non-work wife are not statistically significant. It indicates that the combination situation of work status and intra-household income management pattern might influence the happiness of Japanese married women. It is clear that married women with a successful career and high intra-household bargaining power may feel very happy in Japan.

Fourth, the indexes of absolute income, inter-household relative income, and intra-household relative income are used simultaneously in model 12 to model 16. When other income variables are constant H1 is again supported; the results of inter-household relative income are consistent with those in Tables 7 and 8, therefore the H2 is not supported; the proportion of husband income controlled by the wife positively and significantly affects the wife's happiness. H3 is partly supported again.

5 Conclusions

How does income affect the happiness of married women in Japan? The absolute income hypothesis and relative income happiness are proofed in the published literature. This study develops the relative income hypothesis and makes an empirical study to proof these hypotheses using the Japanese Household Panel Survey (JHPS) data conducted from 1995 and 2013. This study examines the relation between household income, inter-household relative income and intra-household bargaining power and the wife's subjective happiness in Japan based on the pooling OLS, dynamic fix-effect and random effect models.

There are five major conclusions. First, the probability of being happy is higher for the high income group than for the low-income group, therefore H1 [absolute income hypothesis: the happiness of Japanese married women is greater for the higher income (wage) group than for the middle and low income (wage) groups] is supported. The result is consistent with previous studies for other developed countries and developing countries. Second, when absolute income is controlled, the results of inter-household relative income do not support H2 [the happiness of Japanese married women is greater for the group whose income (wage) is higher than for the reference group]; however, when the absolute income is not controlled, H2 is supported. This may be due to the influence of absolute income being greater than the influence of inter-household relative income. Third, the income and education gap between wife and husband can negatively affect happiness; whereas the proportion of husband income controlled by the wife can positively affect happiness, and the affect is greater for the working wife group than the non-work wife group. Thus, H3 [the bargaining power gap between wife and husband affects the Japanese wife's happiness] is partly supported. It is shown that the combination situation of work status and intra-household income management pattern might influence the happiness of Japanese married women. Fourth, the results based on the robustness checks are consistent. Fifth, other factors, such as the wife's education or hours of husband participation in child care or housework can positively affect happiness, and the wife's age, youngest child's age, or living with parents can negatively affect happiness.

Two policy implications emerge from these empirical results. First, the results indicate that the influence of absolute income on happiness is greater when compared with the inter-household relative income. To increase the well-being of married women in Japan, to promote the GDP growth and increase the working income continuously is an important challenge for the Japanese government. From this perspective it is easy to understand why most Japanese have supported the Abe' Mix economic growth promotion policies[13] since 2012.

Second, even though the income/wage gap and education gap between wife and husband negatively affect the wife's happiness, the proportion of husband income controlled by the wife positively affects the wife's happiness, and this influence differs by the non-work wife and working wife groups. It indicates that the traditional gender role consciousness may influence married women's happiness and as a result the work-family conflict for working married women may be severe in Japan. The Japanese government has promoted female labor participation since the 1980s while a progressively ageing population needs to be cared for by someone, and that may well be a working married woman. Even though the government enforces the policies to keep married women working, the proportion of female regular workers is still small, and the female labor participation rate is smaller for Japan than for other developed countries. In order to promote female labor participation in Japan, the implementation of labor and family policies to address the work-family conflict problem is an important challenge for the Japanese government. Importantly, the husband's support with housework and child care may increase the wife's happiness. The implementation and enforcement of a work-life balance policy for both men and women is likely to increase national well-being in Japan.

Finally, there might exist the limitations in this empirical study. For example, even though we utilize the panel data to try to address the heterogeneity problem and initial dependence problem, it is not sufficient to explore the causal effects because the choices or the answers for the independent variables based on the questionnaire are made by respondents themselves, not by random allocation. It is thought that the self-selection bias might influence the estimated results. Thus we should do more researches for the issue to address these bias and problems in the future.

References

Akerlof, G. A., & Yellen, J. K. (1990). The fair wage-effort hypothesis and unemployment. *Quarterly Journal of Economics, 105,* 255–284.

Appleton, A., & Song, L. (2008). Life satisfaction in urban China: Components and determinant. *World Development, 36*(11), 2325–2340.

Boskin, M., & Sheshinski, E. (1978). Optimal redistributive taxation when individual welfare depends upon relative income. *Quarterly Journal of Economics, 92,* 589–601.

[13]Since 2012, the Abe administration has implemented the policy mix of the "three arrows" for reviving the Japanese economy: (1) Aggressive monetary policy; (2) Flexible fiscal policy; and (3) A growth strategy that encourages private sector investment.

Brockmann, H., Delhey, J., Welzel, C., & Yuan, H. (2009). The China puzzle: Falling happiness in a rising Economy. *Journal of Happiness Studies, 10,* 387–405.
Browning, M., Bourguignon, F., Chiappori, P. A., & Lechene, V. (1994). Income and outcomes—A structural model of intrahousehold allocation. *Journal of Political Economy, 102,* 1067–1096.
Browning, M., Chiappori, P. A., & Lewbel, A. (2013). Estimating consumption economies of scale, adult equivalence scales, and household bargaining power. *Review of Economic Studies, 80,* 1267–1303.
Cherchye, L., De Rock, B., Lewbel, A., & Vermeulen, F. (2015). Sharing rule identification for general collective consumption models. *Econometrica, 83,* 2001–2041.
Cherchye, L., De Rock, B., & Vermeulen, F. (2012). Married with children: A collective labor supply model with detailed time use and intrahousehold expenditure information. *American Economic Review, 102,* 3377–3405.
Chiappori, P. A. (1992). Collective labor supply and welfare. *Journal of Political Economy, 100*(3), 437–467.
Chiappori, P. A., Fortin, B., & Lacroix, G. (2002). Marriage market, divorce legislation, and household labor supply. *Journal of Political Economy, 110,* 37–72.
Contoyannis, P., Jones, A., & Rice, N. (2004). The dynamics of health in the British household panel survey. *Journal of Applied Econometrics, 19,* 473–503.
Couprie, H. (2007). Time allocation within the family: Welfare implications of life in a couple. *Economic Journal, 117,* 287–305.
Duesenberry, J. S. (1949). *Income, savings, and the theory of consumer behaviour.* Cambridge: Harvard UP.
Easterlin, R. A. (1974). Does economic growth improve the human lot? In P. A. David & W. B. Melvin (Eds.), *Nations and households in economic growth.* New York: Stanford University Press.
Easterlin, R. A. (2001). Income and happiness: Toward a unified theory. *The Economic Journal, 111,* 465–484.
Ferrer-i-Carbonell, A. (2005). Income and well-being: An empirical analysis of the comparison income effect. *Journal of Public Economics, 89,* 997–1019.
Frank, R. H. (1985). *Choosing the right pond: Human behavior and the quest for status.* Oxford: Oxford University Press.
Hamermesh, D. (1977). Economic aspects of job satisfaction. In O. Ashenfelter & W. Oates (Eds.), *Essays of labor market analysis.* New York: Wiley.
Higuchi, Y., & Hagiwara, R. (2011). Life event and the change of happiness. Keio/Kyoto Global COE Discussion Paper Series, DP2011-016 (in Japanese).
Higuchi, Y., & He, F. (2011). The happiness and life satisfaction of Japanese women. Keio/Kyoto Global COE Discussion Paper Series, DP2011-017 (in Japanese).
Irokawa, T. (1999). Panel data analysis on life status and life satisfaction. *Household Economic Research, 43,* 50–58.
Jiang, S., Lu, M., & Sato, H. (2011). Identity, inequality, and happiness: Evidence from urban China. *World Development, 40*(6), 1190–1200.
Kapteyn, A. B., van Praag, M. S., & van Herwaarden, F. G. (1978). Individual welfare functions and social preference spaces. *Economic Letters, 1,* 173–177.
Layard, R. (1980). Human satisfactions and public policy. *Economic Journal, 90,* 737–750.
Leibenstein, H. (1950). Bandwagon, Snob, and Veblen effects in the theory of consumer's demand. *Quarterly Journal of Economics, 64*(2), 183–207.
Lise, J., & Seitz, S. (2011). Consumption inequality and intra-household allocations. *Review of Economic Studies, 78,* 328–355.
Lise, J., & Yamada, K. (2014). *Household sharing and commitment: Evidence from panel data on individual expenditures and time use* (IFS Working Papers).
Luo, C. (2006). Urban–rural divide, employment, and subjective well-being. *China Economic Quarterly, 5*(3), 817–840. (in Chinese).

Luo, C. (2009). Absolute income, relative income and subjective well-being: Empirical test based on the sample data of urban and rural households in China. *Journal of Finance and Economics, 35*(11), 79–91. (in Chinese).
Ma, X. (2007). Household life time and life inequality. In Y. Higuchi, & M. Seiko (Eds.), *The dynamic of Japanese household behaviour*. Tokyo: Keio University Press (in Japanese).
Ma, X. (2016). Income inequality and subjective happiness in urban region. In H. Kado, & K. Kajitani (Eds.), *Chinese type capitalism going beyond the double trap*. Tokyo: Mineluvi Press (in Japanese).
Otake, F., Shiraishi, S., & Tsutsui, Y. (2010). *Japanese happiness*. Tokyo: Nippon Hyoronsha. (in Japanese).
Sakamoto, K. (2008). Wife's reemployment and intra-household time and consumption allocation on subjective well-being. *Household Economic Research, 77,* 39–51. (in Japanese).
Sano, E., & Otake, F. (2007). Work and happiness. *Journal of Japanese Labor Research, 558,* 4–18. (in Japanese).
Smyth, R., Nielsen, I., & Zhai, Q. (2010). Personal well-being in urban China. *Social Indicators Research, 95,* 231–251.
Tsutsui, Y. (2010). The issue on happiness studies. In F. Otake, S. Shiraishi, & Y. Tsutsui (Eds.), *Japanese happiness*. Tokyo: Nippon Hyoronsha (in Japanese).
Urakawa, K., & Matsuura, T. (2007). Effect of relative income inequality on life satisfaction. *Household Economic Research, 73,* 61–70. (in Japanese).
Vendrik, M. C. M., & Woltjer, G. B. (2007). Happiness and loss aversion: Is utility concave or convex in relative income? *Journal of Public Economics, 91,* 1423–1448.
Wang, P., & VanderWeele, T. J. (2011). Empirical research on factors related to the subjective well-being of Chinese urban residents. *Social Indicators Research, 101,* 447–459.
Wooldridge, J. (2002). *Econometric analysis of cross section and panel data*. Cambridge: The MIT Press.
Wooldridge, J. (2005). Simple solutions to the initial conditions problem in dynamic, nonlinear panel data models with unobserved heterogeneity. *Journal of Applied Econometrics, 20,* 39–54.

Xinxin Ma is an associate professor in Center for Far Eastern Studies (CFES), University of Toyama, Japan. She is a board member of the Japanese Association for Chinese Economy and Management Studies and the editor of *Japanese Journal of Comparative Economics* and *Journal of Chinese Economic and Management*. Her current research project focuses on intra-household bargaining, population ageing and family care in Japan and China. Her articles have appeared in *Journal of Happiness Studies*, *China Economic Review*, *Journal of Chinese Economic and Business Studies*. Her recent book is *Economic Transition and Labor Market Reform in China* (Palgrave Macmillan, 2018) (xxma@eco.u-toyama.ac.jp).

Xiangdan Piao is a research fellow in Departments of Urban and Environmental Engineering, School of engineering, Kyushu University, Japan. She is the member of Japanese Economic Association, American Economic Association, and The Econometric Society. Her current research project focuses on intra-household bargaining power, intra-household allocation inequality, and family relationship in Japan. Her articles have published on *Journal of happiness studies, Economics Bulletin* (piao.xiangdan@doc.kyushu-u.ac.jp).

Time Poverty and Maternal Wellbeing in Japan

Xinxin Ma

Abstract Using the Japanese Maternal Life and Work (LMLW) Survey data conducted in 2012 and 2014, this study explores the determinants of time poverty and its impact on the wellbeing of single mothers and married mothers in Japan. There are six main findings. First, the probability of experiencing time poverty is greater for the longer housework time group and longer working hours group: the time constraint hypothesis is supported. However, housework support from other family members is not a statistically significant influence on time poverty. Second, the influence of household income on the probability of experiencing time poverty is not statistically significant, therefore the income effect hypothesis is not supported. Third, time poverty negatively affects maternal wellbeing. Fourth, the influence of housework hours on the probability of experiencing time poverty is smaller for the single mother group than the married mother group. Fifth, the negative effect of time poverty on subjective health status and mental health disorder is smaller for the single mother group than for the married mother group. Sixth, the negative effect of time poverty on happiness is greater for the single mother group than for the married mother group.

Keywords Time poverty · Wellbeing · Single mother · Married mother · Japan

In Japan, along with the falling birth rate and the increase in the elderly population, the shrinking labor force presents a severe problem. It is thought the aging population may limit economic growth in the long-term.

From the 1980s the Japanese government enforced a set of policies to promote female labor participation in order to address the problem. However, because the typical Japanese husband performs so few housework hours the Japanese wife remains primarily responsible for childcare and housework. As female labor participation increases, based on the time constraint condition, the problem of work-family conflict,[1] sometimes called "double-shift", worsens (Tsuya 2000; Matsuda 2001; Ma

[1]For the study on work-life conflict, please refer Allen et al. (2000), Grzywacz and Brenda (2003), Ford et al. (2007), and Shockley and Singla (2011).

X. Ma (✉)
Center for Far Eastern Studies, University of Toyama, Toyama, Japan
e-mail: xxma@eco.u-toyama.ac.jp

M.-C. Tsai and N. Iwai (eds.), *Quality of Life in Japan*, Quality of Life in Asia 13,
https://doi.org/10.1007/978-981-13-8910-8_5

2007). When housework hours and working hours lengthen, leisure time decreases. Two types of poverty: time poverty and income poverty are distinguished by Deutsch and Siber (2005), Kakwani and Siber (2008), Bardasi and Wodon (2010), Kalenkoski et al. (2011), and Merz and Rathjen (2014). Activity occurring in a lifetime is classified into three types: committed activity (working, commuting, childcare and family care), necessary activity (sleep, eating, going to the toilet), and other activity (hobby, travel). Other activity is usually defined as discretionary activity. The time poverty line is defined as being at 50% of the median of the discretionary activity time. When the discretionary activity time is less than the time poverty line, it is defined as time poverty. It is thought for Japanese women time poverty may result from long working hours and long housework hours. Many studies analyze income poverty in Japan but there are few empirical studies of time poverty. This study aims to address this gap in knowledge.

The issue is of importance because most single-mother households fall more easily into income poverty than married mother households and the figures are disturbingly high.[2] In order to address the problem the Japanese government enforced a policy to encourage the single mother group to enter the labor market. These labor policies may reduce economic poverty for the single mother group but the single mother group is less likely to receive housework support from a husband than the married mother group (households with wife and husband), therefore the probability of experiencing time poverty is greater for the single mother group than for the married mother group. This study also compares the single mother group and the married mother group on the issue of housework support from a partner. As a quasi-natural experiment the comparison study can give us more evidence about the influence of husband housework support on maternal time poverty in Japan.

It is thought that time poverty may reduce individual utility or wellbeing. Kalenkoski et al. (2011) was the first published study to investigate the influence of time poverty on subjective health status. Yet empirical studies on the relation between time poverty and wellbeing are scare. This study will investigate the influence of time poverty on maternal wellbeing in Japan and give better evidential context for the development of poverty reduction policies. It should be noted that Kalenkoski et al. (2011) only utilized one kind of wellbeing index. This study utilizes three wellbeing indexes: the subjective health status, mental health disorder, and subjective happiness.

Using the Japanese Maternal Life and Work (JMLW) Survey data conducted in 2012 and 2014 (JMLW2012 and 2014), this study investigates the following questions. In Japan, (1) what determines time poverty? Do working hours/housework hours (time constraint), and household income affect the probability of experiencing time poverty? (2) Does time poverty affect maternal wellbeing? (3) Are the determinants of time poverty and its influence on wellbeing different for the single mother

[2]Based on the data of the National Life Basic Survey conducted in 2014 by Ministry of Health, Labor and Welfare Japan, it is shown that the child relative poverty rate was 16.3% in 2013. It indicates that one in six children is in poverty. In addition, the child relative poverty rate was 54.6% for a one-parent household with a child aged less than 16 in 2013.

group and the married mother group? For the issue (1), the time constraints hypothesis (Hypothesis 1) and household income effect hypothesis (Hypothesis 2) are tested; for the issue (2), it is expected that the time poverty may negatively affect wellbeing (Hypothesis 3); for the issue (3), it is thought that, the influence of time constraint and income effect on the probability of experiencing time poverty may differ by the single mother group and the married mother group, and the influence of time poverty on maternal wellbeing also may differ by these two groups.

The remainder of this chapter is as follows. Section 1 summarizes the economic theory on time poverty and its influence on wellbeing, and previous empirical studies on these issues and sets the three hypotheses in this context. Section 2 explains the methods of analysis, including introduction to models and data. Sections 3 and 4 give the calculated results. The next section states and interprets the econometric results. Section 5 presents the main findings and policy implications.

1 Literature Review and Hypotheses Setting

The first question is, what factors determine time poverty? Based on household consumption and utility theory, because the individual or household faces income and time constraints, the working hours, housework hours and income may affect the probability of experiencing time poverty (Becker 1965; Gronau 1977).

First, based on the time constraint condition, it is thought that the longer the housework hours and working hours are, the greater the probability of time poverty occurring. The probability of experiencing time poverty may be smaller for the group who receive more childcare or housework support from other family members (H1).

Second, when the household can buy childcare or housework services from the market (i.e. baby helper, housework robot), the probability of experiencing time poverty may decrease. The probability of experiencing time poverty may be smaller for high-income households than for low-income households (this is named income effect in this study). However, when the working hours are much longer for the high-income household, the income effect on time poverty may be not significant. Thus, even though it is thought income may affect time poverty, the income effect direction (negative effect or positive effect) is not clear and should be investigated further using the empirical study results (H2).

Turning to the empirical studies on the issue, Bardasi and Wodon (2010), Kalenkoski et al. (2011), Merz and Rathjen (2014), and Ishii and Urakawa (2014) analyze individual characteristics (age, gender, race), family factors (married status, family member, child number, child age), labor market factor (work status, occupation), human capital (education, health) and regions on time poverty but the results are not consistent. Particularly, these studies do not focus on H1 and H2 tests. This study focuses on the time constraint and income effect on time poverty. This study constructs a set of variables for time constraint to investigate the influence of working hours, housework hours and family member support on time poverty. These new findings supply more evidence for the time poverty issue.

H1 (time constraint hypothesis)
In Japan the probability of experiencing time poverty is greater for the group with longer working hours and longer housework hours; the probability of experiencing time poverty is smaller for the group who receive more childcare or housework support from other family members.

H2 (income effect hypothesis)
In Japan household income affects the probability of experiencing time poverty.

The second issue concerns the influence of lifetime allocation (working hours, housework hours) on individual wellbeing. The published studies confirm wellbeing is a complex concept. Health status, happiness, or life-satisfaction are usually utilized as the index of wellbeing. On the issue of the influence of working hours on mental health, the previous studies found that the probability of becoming a patient with mental health disorder is greater for workers with longer working hours (Liff 1981 for the USA; Perrucci et al. 2007 for USA and European countries; Tuttle and Garr 2009 for European countries; Ma 2009 for Japan). Kalenkoski et al. (2011) indicate that the subjective health status is better for the time poverty group. Thus these previous empirical study results are not consistent. Based on utility theory, previous studies and the time poverty situation in Japan, this study will proof the H3 as follows.

H3 (time poverty harms maternal well-being hypothesis)
Time poverty negatively affects maternal wellbeing in Japan.

The third issue considers the difference between the single mother group and the married mother group. It is thought that when mothers receive more housework support from other family members (husband, parents), the probability of experiencing time poverty may become smaller. It is pointed out that the parents' support for housework and childcare is slight in Japan (Tsuya 2000; Matsuda 2001; Ma 2007). However, when parent support is greater for the single-mother group than the married mother group, the probability of experiencing time poverty may be smaller for the single-mother group. The difference in the influence of time poverty on wellbeing between these two groups is not clear. This paper employs a comparison study of these two groups and presents new finding for the time poverty issue (Fig. 1).

2 Methodology and Data

2.1 Data

This study utilizes the Japanese Maternal Life and Work Survey (JMLW) data. The survey was conducted in 2012 and 2014 (JMLW2012 and 2014) by the Japanese Institute of Labor Policy and Training (JILPT). The sampling method is a stratified random sampling based on samples from the basic residential register. The survey covers 175 regions across Japan. The survey objective is single mother households and married mother households with a child aged less than 18 years old on November

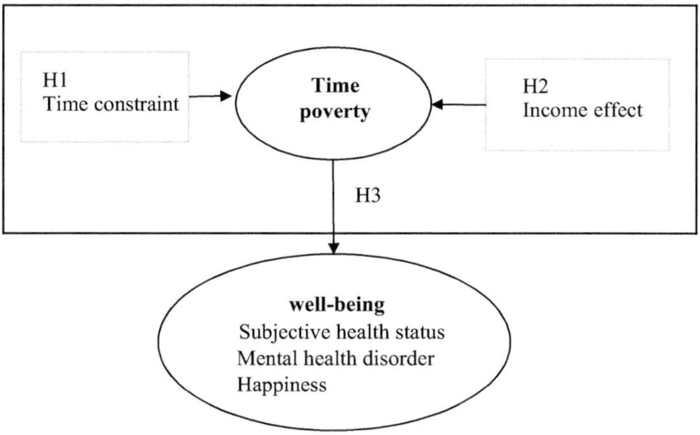

Fig. 1 Empirical study framework. *Source* by the author

1st, 2012 and 2014. These are two period cross-section surveys. The surveys are conducted by interview and indwelling recovery method. The collected sample consists of 2,200 in 2012, and 2,198 in 2014. Missing value samples are deleted after the independent and dependent variables are constructed. The main advantages of using the survey data are as follows: first, the survey covers all of Japan and the sampling is random, therefore the survey data portrays comprehensive maternal information for Japan. Second, the survey questionnaire gathers information on maternal individual characteristics, work status, children, parents, husbands, household income and wellbeing. This data can be utilized to proof H1, H2 and H3. Third, to maintain the samples to employ the analysis for the single mother group, the collected samples for the single mother groups in JMLW2012 and 2014 are more than the national actual values. Therefore, the survey data can be utilized to compare the single mother group and the married mother group. The comparison study of these two groups provides the first evidence of this kind for the time poverty issue.

2.2 *Model*

This study employs a set of models to analyze H1, H2, and H3. First, to analyze the determinants of time poverty, a probit regression model is utilized. This is shown in Eq. (1).

$$\begin{aligned} P(Y_i = 1) = p_i = P(v_i = 1 - a - \beta_T \text{ Time}_i - \beta_I \text{ Income}_i - \beta_X X_i) \\ Y_i^* = a + \beta_T \text{ Time}_i + \beta_I \text{ Income}_i + \beta_X X_i + v_i \\ Y_i^* = 1 \text{ if } Y_i^* > 0 \\ Y_i^* = 0 \text{ if } Y_i^* < 0 \end{aligned} \quad (1)$$

In Eq. (1), $P(Y_i = 1)$ is the independent variable of the probability of experiencing time poverty, it is a binary variable (equal to 1 if the women is in time poverty, and equal to 0 if not). Y^* is a latent variable of probability of experiencing time poverty. i stands for the individual, a is the constant term, and v is the error term. Time represents a time constraint variable (work status, children number, youngest child age, living with parents, husband housework support), Income represents equivalent household income, X represents the other control variables. β_T, β_I, and β_X are the estimated coefficient for each variable. When β_T, β_I, are statistically significant, it is shown that time constraint factors and the household income affect the probability of experiencing time poverty, therefore H1 and H2 are supported.

Second, to investigate the influence of time poverty on wellbeing, the following models are utilized. (1) The ordered logit regression model is utilized to analyze subjective health status. (2) The mental health score is calculated and OLS (Ordinary Least Squares) estimation is employed to analyze mental health status. (3) The variable of subjective happiness is a value from 1 (the most unhappiness) to 10 (the most happiness), and two models-OLS and the ordered logit model can be utilized to analyze subjective happiness. The estimated results based on OLS are summarized because OLS estimation is easily understood and explained. (4) To address the endogenous problem, a two-stage estimation model is utilized. In the first step, the imputed probability of experiencing time poverty is calculated based on the time poverty probability function; in the second step, the imputed time poverty values are used in wellbeing functions. The health and household income quintile dummy variables in the time poverty probability function are used to be identify variables. When the coefficients of imputed time poverty probability variables are significant statistically, it means that time poverty influences the wellbeing of Japanese women.

$$\mathrm{Wellbeing}_i = b + \gamma_{TP}\mathrm{TP}_i^* + \gamma_H H_i + u_i \tag{2}$$

In Eq. (1), Wellbeing is the independent variable of three sorts of wellbeing index (subjective health status, mental health score which represents mental health disorder, and happiness). i stands for the individual, b is the constant term, and u is the error term. TP^* represents the imputed time poverty variable which is calculated based on the results by Eq. (1). H represents the other control variables. γ_{TP}, and γ_H are estimated coefficient for each variable. When γ_{TP} is statistically significant, it is found that time poverty affects wellbeing, therefore H3 is supported.

2.3 *Variable Setting*

In relation to the probability function of time poverty, time poverty is defined as follows. In previous studies lifetime activity is classified into three types: committed activity (working, commuting, childcare and family care), necessary activity (sleeping, eating, going to the toilet), and other activity (hobby, travel). Other activity is usually defined as discretionary activity. Thus discretionary activity time is calculated as the value of 24 h daily excepted both committed activity time and necessary activity time daily. However, it is thought that necessary activity time is a part of individual leisure hours when the committed activity time becomes longer (working hours, childcare and family care hours becoming longer) the individual can adjust necessary activity time (e.g. reduce sleep hours). Thus in this study, in order to explore the effect of working hours and childcare hours (committed activity time) on wellbeing, discretionary activity time is composed of the other time and necessary activity time, which is calculated as the value of 24 h daily excepted committed activity time daily. The time poverty line is defined as being at 50% of the median of discretionary activity time. When the discretionary activity time is less than time poverty line, it is defined as time poverty. The dependent variable is a binary variable that is equal to 1 when the individual experiences time poverty and equal to 0 when the individual experiences non-time poverty.

For the wellbeing function, the independent variables are constructed as follows. First, there are five answer options in the subjective health status function: 1 Very poor health, 2 Poor health, 3 Fair, 4 Healthy, and 5 Very healthy. The ordered category variable is utilized as the dependent variable. Second, in the mental health function, the mental health score is calculated and utilized as a dependent variable. There are eight answer options for mental health disorder: "I usually feel troubled by something"; "I cannot concentrate on things"; "I am depressed"; "It is too much trouble to do anything"; "I feel scared"; "I cannot sleep"; "I feel lonely"; "No motivation to do anything". There are five answer options for frequency: "No", "1–2 days", "3–4 days", "more than 5 days" weekly. The scores are set from 0 to 3. It is scored as 0 when the individual selects "No", and as 3 when the individual selects "more than 5 days". The total value of these score is the mental health score. The probability of experiencing a mental health disorder is higher for the individual with a higher mental health score. Lastly, in the function of subjective happiness, the dependent variable is a value from 1 to 10 based on the answer option for the question "Did you feel happy over the last year?". "Very happy" is valued as "10", and "Very unhappy" is valued as "0".

Dependent variables are the variables likely to affect time poverty, and wellbeing. First, for the index of time constraint hypothesis, (1) work status is utilized as the market work index to prove H1. Work status is a set of dummy variables: they are non-work, self-employed, regular worker and irregular worker. When the working hours are longer (for example, work as a regular worker), the probability of experiencing time poverty may be greater because time is constrained by various activities. H1 is supported when the coefficient of the regular worker is statistically significant and the value is positive. (2) The youngest child's age and number of children are utilized as the housework hours index. The youngest child's age dummy variables are aged 0–3, 4–6, and over 7 years old. The child number dummy variables are set as one child, two children, and more than three children. It is thought that when the child is younger and the number of children is greater the housework hours may increase, thus these variables may influence the probability of experiencing time poverty. H1 is supported when the coefficients of these variables are statistically significant. (3) Living with parents,[3] and the proportion of husband housework hours dummy variables are used as a set of family support factors. The proportion of husband housework hours to total housework hours are categorized as 0, 10, 20, 30, 40, and more than 50%. The proportion of husband housework hours to total housework hours is counted as 0% for the single mother group. It is expected that childcare support and husband housework participation will reduce maternal time poverty. H1 is supported when the coefficients of these variables are statistically significant and the values are positive.

Second, based on equivalent household income, income from first to fifth quintile dummy variables is utilized. H2 is supported when coefficients of the high-income group are statistically significant and the values are negative. Lastly, other factors may influence the probability of experiencing time poverty and well-being, the individual characteristic variables such as education dummy variables (junior high school, senior high school, vocational school, college, university), age, age squared, health dummy variables (equal to 1 if "Very healthy" and "Healthy"; equal to 0 if "Very poor health", "Poor health", and "Fair") are utilized as an index of human capital. In addition, regional dummy variables (Decree city, City with population more than 200 thousand, City with population more than 100 thousand and less than 200 thousand, City with population less than 100 thousand, and Towns and villages), and year dummy are also utilized to control the macroeconomic environment influence.

The statistical description of dependent variables and independent variables is summarized in Table 1. Totals for (total samples of single mother group and married mother group), the single mother group and the married mother group are summarized separately. First, the t-test is used to compare the mean value difference between the single mother group and the married mother group. It is found that the group characteristics differ by these two groups. For example, the proportion of high education group (college, university) is smaller, whereas the proportion of low education group (junior high school, senior high school) is greater for the single mother group than for the married mother group, the t-test coefficients are significant statistically for these education variables.

Second, the percentage for the single mother group is 22.4%. This is more than the actual proportion of the single mother group in the whole population. This empirical study cannot be used for the single mother group when the survey sample is based on the actual proportion of single mothers because the proportion of single mothers is so small. The total sample (the total of single mother group and married group) it is adjusted by using a reweight method.

Third, the summary should also consider the characteristic difference between the analyzed sample and survey sample because the missing value sample is deleted in the analysis. Table 2 summarizes the observations and mean values of these two samples. The difference of sample, the mean value between these two groups, is also calculated. Approximately twenty percent of the sample is not used by the missing value, the missing sample by no-answer is the husband housework support variable and it is greatest. When the mean values of the analyzed sample and the survey

[3]Living with a parent is defined as the woman living with her parents or her husband's parents in the same house.

Table 1 Descriptive statistics of analyzed samples

	Total		Single mother		Married mother		(4) t-test
	Mean	S.D.	Mean	S.D.	Mean	S.D.	t-value
Time poverty	0.034	0.182	0.018	0.134	0.039	0.193	3.191**
Mental health	11.728	4.596	13.104	5.418	11.331	4.249	−9.198***
Subjective health status	2.272	1.065	2.557	1.100	2.190	1.040	−11.082***
Happiness	6.931	2.319	6.054	2.513	7.182	2.197	14.102***
Single mother	0.224	0.417	1.000	0.000	0.000	0.000	
Age	41.128	6.701	40.848	6.620	41.208	6.724	0.651*
Education							
Junior high school	0.050	0.217	0.104	0.306	0.034	0.181	−9.371***
Senior high school	0.379	0.485	0.459	0.499	0.356	0.479	−5.938***
Vocational school	0.157	0.364	0.148	0.355	0.160	0.366	0.899
College	0.243	0.429	0.187	0.390	0.259	0.438	3.559***
University	0.171	0.377	0.103	0.304	0.191	0.393	9.041***
Health	0.560	0.496	0.461	0.499	0.588	0.492	8.297***
Work status							
Regular worker	0.248	0.432	0.370	0.483	0.213	0.409	−7.091***
Irregular worker	0.396	0.489	0.464	0.499	0.377	0.485	−6.955***
Self-employed	0.080	0.271	0.048	0.214	0.089	0.285	5.061***
Non-work	0.276	0.447	0.118	0.323	0.321	0.467	12.233***
Child number	2.011	0.826	1.799	0.814	2.072	0.820	8.994***
Youngest child age							
Aged 0–3	0.170	0.375	0.052	0.222	0.203	0.403	11.664***
Aged 4–6	0.170	0.376	0.125	0.331	0.183	0.387	4.784***

(continued)

Table 1 (continued)

	Total		Single mother		Married mother		(4) t-test
	Mean	S.D.	Mean	S.D.	Mean	S.D.	t-value
Aged 7 and older	0.661	0.474	0.823	0.382	0.614	0.487	−12.866***
Living with parents	0.234	0.423	0.329	0.470	0.207	0.405	−7.217***
Husband housework support	1.753	1.712	0.319	1.066	2.167	1.638	30.020***
Household income quintiles							
1st	0.213	0.409	0.253	0.435	0.201	0.401	−6.206***
2nd	0.131	0.338	0.322	0.468	0.076	0.265	−19.279***
3rd	0.207	0.405	0.217	0.412	0.204	0.403	0.200
4th	0.224	0.417	0.118	0.323	0.254	0.435	9.819***
5th	0.226	0.418	0.090	0.286	0.265	0.441	13.178***
Regions							
Decree city	0.123	0.329	0.146	0.354	0.117	0.321	−1.101
Population more than 200 thousands	0.284	0.451	0.267	0.443	0.288	0.453	1.061
Population more than 100 thousands and less than 200 thousands	0.197	0.398	0.160	0.367	0.207	0.406	2.623**
Population less than 100 thousands	0.148	0.355	0.166	0.372	0.142	0.349	−2.020**
Towns and villages	0.249	0.432	0.260	0.439	0.245	0.430	−0.979
Survey year							
2014	0.478	0.500	0.512	0.500	0.468	0.499	−3.405***
2014	0.522	0.500	0.488	0.500	0.532	0.499	3.405***
Observations	3179		711		2468		

Note
1. t-test for variables means of married mother group and single mother group
2. *, **, *** Statistical significant levels are 10, 5, 1%
Source calculated based on 2012–2014 Japanese Maternal Life and Work (JMLW) survey data

Table 2 Comparison of descriptive statistics of analyzed sample and survey sample

	(1) Survey sample			(2) Analyzed sample			D = (2) − (1)		
	Obs.	Mean	S.D.	Obs.	Mean	S.D.	Obs.	Mean	S.D.
Time poverty	3179	0.034	0.182	4398	0.030	0.171	1219	−0.004	−0.011
Mental health	3179	11.728	4.596	4398	11.427	5.247	1219	−0.301	0.651
Subjective health status	3179	2.272	1.065	4229	2.330	1.079	1050	0.058	0.014
Happiness	3179	6.931	2.319	4147	6.752	2.404	968	−0.179	0.085
Single mother	3179	0.224	0.417	4397	0.306	0.461	1218	0.082	0.044
Age	3179	41.128	6.701	4393	41.331	6.936	1214	0.203	0.235
Education									
Junior high school	3179	0.050	0.217	4149	0.064	0.244	970	0.014	0.027
Senior high school	3179	0.379	0.485	4149	0.391	0.488	970	0.012	0.003
Vocational school	3179	0.157	0.364	4149	0.152	0.359	970	−0.005	−0.005
College	3179	0.243	0.429	4149	0.227	0.419	970	−0.016	−0.010
University	3179	0.171	0.377	4149	0.167	0.373	970	−0.004	−0.004
Health	3179	0.560	0.496	4229	0.540	0.498	1050	−0.020	0.002
Work status									
Regular worker	3179	0.248	0.432	4250	0.283	0.451	1071	0.035	0.019
Irregular worker	3179	0.396	0.489	4250	0.390	0.488	1071	−0.006	−0.001
Self-employed	3179	0.080	0.271	4250	0.080	0.271	1071	0.000	0.000
non-work	3179	0.276	0.447	4250	0.247	0.431	1071	−0.029	−0.016
Child number	3179	2.011	0.826	4286	1.996	0.856	1107	−0.015	0.030

(continued)

Table 2 (continued)

	(1) Survey sample			(2) Analyzed sample			D = (2) − (1)		
	Obs.	Mean	S.D.	Obs.	Mean	S.D.	Obs.	Mean	S.D.
Youngest child age									
Aged 0–3	3179	0.170	0.375	4398	0.148	0.355	1219	−0.022	−0.020
Aged 4–6	3179	0.170	0.376	4398	0.154	0.361	1219	−0.016	−0.015
Aged 7 and older	3179	0.661	0.474	4398	0.698	0.459	1219	0.037	−0.015
Living with parents	3179	0.234	0.423	4398	0.239	0.427	1219	0.005	0.004
Husband housework support	3179	1.753	1.712	3499	1.751	1.710	320	−0.002	−0.002
Household income quintiles									
1st	3179	0.213	0.409	4208	0.267	0.442	1029	0.054	0.033
2nd	3179	0.131	0.338	4208	0.139	0.346	1029	0.008	0.008
3rd	3179	0.207	0.405	4208	0.195	0.397	1029	−0.012	−0.008
4th	3179	0.224	0.417	4208	0.200	0.400	1029	−0.024	−0.017
5th	3179	0.226	0.418	4208	0.199	0.399	1029	−0.027	−0.019
Regions									
Decree city	3179	0.123	0.329	4398	0.126	0.332	1219	0.003	0.003
Population more than 200 thousands	3179	0.284	0.451	4398	0.280	0.449	1219	−0.004	−0.002
Population more than 100 thousands and less than 200 thousands	3179	0.197	0.398	4398	0.195	0.396	1219	−0.002	−0.002
Population less than 100 thousands	3179	0.148	0.355	4398	0.154	0.361	1219	0.006	0.006
Towns and villages	3179	0.249	0.432	4398	0.245	0.430	1219	−0.004	−0.002
Survey year									
2014	3179	0.478	0.500	4398	0.500	0.500	1219	0.022	0.000
2014	3179	0.522	0.500	4398	0.500	0.500	1219	−0.022	0.000

Source calculated based on 2012–2014 Japanese Maternal Life and Work (JMLW) survey data

sample are compared, the difference of variable mean values (see D = (2) − (1)) is smaller. It is thought the influence of missing samples on analyzed results might be smaller.

3 Descriptive Statistics Results of Time Poverty

3.1 Allocation of Lifetime Activity by the Single Mother and Married Mother Groups

The average daily lifetime activity for the single mother group and for the married mother group is summarized in Table 3.

First, working hours is longer for the single mother group (5.29 h) than for the married mother groups (4.43 h). Travel to work time is also longer for the single mother group (0.37 h) than for the married mother groups (0.34 h), but the differentials of travel to work time is quite small between these two groups.

Second, both childcare time and housework time are shorter for the single mother group than childcare time and housework time for the married mother group. For example, childcare time is 3.82 h for the single mother group, whereas it is 4.90 h for the married mother group; housework time is 2.28 h for the single mother group, whereas it is 3.53 h for the married mother group.

Third, sleep time is shorter for the single mother group (5.75 h) than for the married mother group (6.12 h).

These results show that working time is longer, childcare and housework time is shorter, and sleep time is shorter for the single mother group than for the married mother group. When market work preference and leisure preference are similar, it is shown that the quality of life is lower for the single mother group than for the married mother group.

Table 3 Allocation of average daily time by single mother and married mother groups (Unit: Daily hours)

		Single mother		Married mother	
		Mean	S.D.	Mean	S.D.
a	Working hours	5.292	1.868	4.427	2.112
b	Commuting hours	0.367	0.296	0.344	0.290
c	Child care hours	3.822	2.037	4.896	2.074
d	Housework hours	2.277	1.531	3.526	2.112
	Total (a + b + c + d)	11.758		13.192	
	Sleep hours	5.748	1.623	6.125	1.510

Source calculated based on 2012–2014 Japanese Maternal Life and Work (JMLW) survey data

3.2 *Kernel Density Distribution of Working Hours and Household Income*

Kernel density distribution of working hours and annual equivalent household income (tax included) is shown in Fig. 2.

First, the density of samples in which the working hours are 0 h is more for the married mother group than for the single mother group, whereas the density of samples for which working time is longer is more for the single mother group.

Second, the density of samples whose annual equivalent household income is 0 Yen is more for the single mother group, whereas the density of samples with a high-household income is more for the married mother group. The calculated average equivalent household income (included tax, Japanese yen) is 2,140 thousand for the single mother group, and 5,140 thousand for the married mother group. The calculated average equivalent household income (excluded tax, Japanese yen) is 1,300 thousand for the single mother group, and 3,120 thousand for the married mother group.

These results indicate that the working time is longer and the household income is lower for the single mother group than for the married mother group. The probability of time poverty and income poverty is greater for the single mother group and this may cause the quality of life to be lower for the single mother group than for the married mother group.

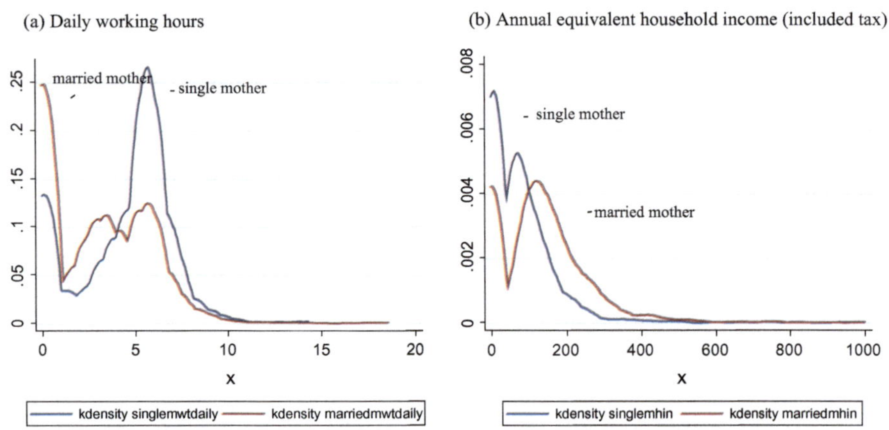

Fig. 2 Kernel density distribution of working hours and household equivalent income. *Source* calculated based on 2012–2014 Japanese Maternal Life and Work (JMLW) survey data. *Note* red line: married mother; bule line: single mother

4 Econometric Analysis Results

4.1 Determinants of Time Poverty: Testing for H1, H2

For issue one, the probability of experiencing time poverty is analyzed using a probit regression model in order to investigate the determinants of time poverty and to test the H1 and H2. The results are summarized in Table 4 by (1) total group, (2) single mother group and (3) married mother group. The test results for H1 and H2 are discussed based on the results for the total group. There are three main findings.

First, regarding the influence of working hours (H1), (1), the probability of experiencing time poverty is greater for the worker group (both self-employed worker, irregular worker, and regular worker) than for the non-worker group. The coefficients of the self-employed, irregular worker, and regular worker dummy variable are statistically significant. They are 0.697 for a self-employed worker, 0.617 for a regular worker, and 0.416 for an irregular worker. The reasons for these results are worthy of discussion. Working hours are longer for a regular worker (e.g. a full-time worker) than for an irregular worker (e.g. a part-time worker) therefore it is not surprising that the probability of experiencing time poverty is greater for regular workers. In addition, even though it is thought that a self-employed worker can adjust working hours by his (her) self, when the working hours are longer for the self-employed worker, the probability of experiencing time poverty might be greater for the self-employed worker than the employee group.

In respect of (2), regarding the influence of childcare and housework time, the probability of experiencing time poverty is greater for the group with the youngest child aged 0–3 years old than for the group with youngest child aged more than 4 years old. In addition, the probability of experiencing time poverty is greater for the group with more children at the statistical 1% level. These results indicate that the probability of experiencing time poverty is greater for the group with longer childcare and housework time. In respect of (3), to look at the influence of family member support factors, the proportion of husband's housework hours, living with parents dummy variables are not statistically significant. Therefore, the first half of H1 (the probability of experiencing time poverty is greater for the group with longer working hours and longer housework hours) is supported, but the latter half of H1 (whereas the probability of experiencing time poverty is smaller for the group who receive more childcare or housework support from other family members) is not supported.

Second, to look at the income effect on time poverty (H2), the whole coefficients of the income quintile dummy variables are not statistically significant. Thus the H2 (the probability of experiencing time poverty is smaller for the high-income group) is not supported. The results can be explained as the follows. It is thought that there are two kinds of effects (positive effect and negative effect) which can explain the influence of household income on time poverty. The first is the market service consumption effect. A high-income household can utilize the childcare or housework service by buying from the market (positive effect). For example, they

Table 4 Results of probability of experiencing time poverty

	(1) Total		(2) Single mother		(3) Married mother	
	coeff.	z-value	coeff.	z-value	coeff.	z-value
Single mother	−0.233	−1.55				
Age	0.056	0.82	−0.125	−0.47	0.056	0.74
Age squared	−0.001	−0.78	0.002	0.64	−0.001	−0.8
Education (ref.: Junior high school)						
Senior high school	−0.411**	−2.02	−1.088**	−2.31	−0.305	−1.21
Vocational school	−0.180	−0.84	−0.884*	−1.6	−0.077	−0.29
College	−0.269	−1.28	−0.565	−1.21	−0.199	−0.78
University	−0.377*	−1.69	−1.080	−1.56	−0.280	−1.05
Health	−0.080	−0.87	0.701**	2.07	−0.180*	−1.79
Work status (ref.: non-work)						
Regular worker	0.617***	4.38	4.733	0.01	0.556***	3.77
Irregular worker	0.416***	3.08	4.010	0.01	0.444***	3.19
Self-employed	0.697***	4.09	5.560	0.01	0.652***	3.63
Child number	0.152***	2.74	0.112	0.54	0.184***	3.04
Youngest child age (ref.: aged 0–3)						
Aged 4–6	−0.258*	−1.74	4.167	0.00	−0.253*	−1.63
Aged more than 7	−0.481***	−3.2	4.422	0.00	−0.513***	−3.18
Living with parents	0.073	0.68	−0.147	−0.46	0.133	1.12
Husband housework support	0.021	0.76	0.115	1.17	0.015	0.51
Household income (ref.: 1st quintile)						
2nd	−0.112	−0.61	1.013*	1.73	−0.367	−1.55
3rd	0.056	0.39	0.873	1.45	−0.054	−0.35
4th	0.102	0.72	0.965	1.53	0.018	0.12
5th	0.151	1.03	–	–	0.175	1.15
Regions	Yes		Yes		Yes	
Years	Yes		Yes		Yes	
Constant	−2.979**	−2.16	−9.578	−0.01	−2.865**	−1.89
Observations	3179		711		2468	
Log likelihood	−440.171		−46.404		−377.219	
R2	0.0731		0.2711		0.0701	

Note *, **, *** Statistical significant levels are 10, 5, 1%
Source calculated based on 2012–2014 Japanese Maternal Life and Work (JMLW) survey data

can utilize public or private childcare services, and buy food and cleaning services. These market services may reduce time poverty for the high-income household. The second is the time constraint effect. It is thought that working hours may be longer for a high-income household, thus the discretionary activity time may be shorter for the high-income household than for the low-income household (negative effect). It is thought the income effect of household income is the total results of the positive and negative effects. The results indicate that the positive effect is less than the negative effect. These results indicate that the market service consumption effect (positive effect) is smaller than the time constraint effect (negative effect). The reasons can be considered as follows. First, maternal childcare cannot be substituted by market childcare services. Second, the price of market services is so high that even middle and high-income households cannot afford these services. For example, the wage of a childcare helper and a housework helper is high in Japan. Thus, even the high-household income group cannot solve the time poverty problem.

Third, the influences of other factors on time poverty are as follows. (1) In comparison with the low education group (junior high school), the probability of experiencing time poverty is smaller for middle-education groups (senior high school) and high-education groups (college and university), these results are consistent with Bardasi and Wodon (2010), Kalenkoski et al. (2011), Merz and Rathjen (2014), and Ishii and Urakawa (2014). (2) The coefficient of maternal age is not statistically significant. It might be because that the maternal age effect on time poverty is subsumed in the effects of child number and the youngest child age.

4.2 Influence of Time Poverty on Maternal Wellbeing: H3 Testing

The results of the second issue, the influence of time poverty on wellbeing, are summarized in Table 5 (subjective health status), Table 6 (mental health score) and Table 7 (subjective happiness).

The results using total sample (1) are used to proof H3. First, for the estimation on subjective health status (Table 5), the coefficient of the time poverty dummy variable is −11.293, which is statistically significant at the 1% level. Second, for the estimation on subjective health status, the coefficient of the income poverty dummy variable is 6.339, which is statistically significant at the 1% level (Table 6). Third, for the estimation of subjective happiness, the coefficient of time poverty is −2.782, which is also statistically significant at the 1% level (Table 7). It is shown that the probability of experiencing poor health status and mental health disorder is greater, and the probability of feeling happiness is smaller when the mother is in time poverty. Thus, the H3 (time poverty negatively affects wellbeing in Japan) is supported.

Table 5 Results of the influence of time poverty on subjective health status

	(1) Total		(2) Single mother		(3) Married mother	
	coeff.	z-value	coeff.	z-value	coeff.	z-value
Imputed time poverty	−11.293***	−22.49	−6.287***	−8.85	−18.775***	−20.63
Single mother	−3.343***	−20.06				
Age	0.572***	10.18	0.190*	1.66	1.017***	13.29
Age squared	−0.007***	−10.55	−0.003*	−1.85	−0.012***	−13.54
Education (ref.: Junior high school)						
Senior high school	−3.736***	−14.55	−1.940***	−5.07	−6.593***	−15.10
Vocational school	−1.027***	−5.27	−0.190	−0.60	−2.252***	−7.65
College	−1.804***	−8.49	−0.859**	−2.56	−3.687***	−10.77
University	−3.002***	−11.78	−1.310***	−3.17	−5.706***	−13.49
Work status (ref.: non-work)						
Regular worker	7.234***	22.2	5.093***	10.06	11.625***	20.38
Irregular worker	4.898***	22.09	3.348***	9.61	8.038***	20.59
Self-employed	8.153***	21.87	5.480***	8.73	13.211***	20.49
Child number	1.724***	20.25	0.830***	6.02	2.891***	19.82
Youngest child age (ref.: aged 0–3)						
Aged 4–6	−3.016***	−17.21	−2.313***	−5.63	−4.933***	−18.32
Aged more than 7	−5.609***	−21.07	−3.517***	−7.32	−9.169***	−20.17
Living with parents	0.950***	10.82	0.939***	5.84	1.354***	11.52
Husband housework support	0.310***	12.58	0.287***	4.11	0.457***	15.00
Household income	0.012***	15.21	0.006***	3.42	0.018**	17.00
Household income-sq.	−1.130E−05***	−8.87	−2.050E-06		−1.950E-05***	−11.33
Regions	Yes		Yes		Yes	
Years	Yes		Yes		Yes	
Observations	3179		711		2468	
Log likelihood	−3961.182		−942.362		−2915.052	
R2	0.098		0.088		0.123	

Note *, **, *** Statistical significant levels are 10, 5, 1%
Source calculated based on 2012–2014 Japanese Maternal Life and Work (JMLW) survey data

Table 6 Results of the influence of time poverty on mental health

	(1) Total		(2) Single mother		(3) Married mother	
	coeff.	z-value	coeff.	z-value	coeff.	z-value
Imputed time poverty	6.339***	6.18	4.719**	2.40	7.031***	5.62
Single mother	3.007***	8.37				
Age	−0.639***	−5.05	−0.969***	−3.04	−0.562***	−4.03
Age squared	0.007***	4.87	0.011***	2.96	0.006***	3.92
Education (ref.: Junior high school)						
Senior high school	0.938*	1.63	−0.266	−0.24	1.886***	2.68
Vocational school	−0.293	−0.64	−0.826	−0.92	0.457	0.82
College	−0.307	−0.63	−0.492	−0.51	0.409	0.68
University	0.610	1.06	−0.361	−0.31	1.435**	2.07
Work status (ref.: non-work)						
Regular worker	−4.865***	−7.22	−6.470***	−4.75	−4.836***	−5.98
Irregular worker	−3.478***	−7.54	−4.801***	−5.08	−3.550***	−6.35
Self-employed	−4.752***	−6.15	−6.306***	−3.69	−4.883***	−5.33
Child number	−1.074***	−6.04	−0.770**	−2.02	−1.223***	−5.87
Youngest child age (ref.: aged 0–3)						
aged 4–6	2.240***	5.87	4.451***	3.90	2.053***	4.82
aged more than 7	4.153***	7.47	5.080***	3.81	4.219***	6.44
Living with parents	−0.842***	−4.16	−1.193***	−2.67	−0.702***	−3.07
Husband housework support	−0.253***	−4.46	−0.167	−0.87	−0.287***	−4.94
Household income	−0.006***	−3.41	−0.005	−0.98	−0.006***	−3.31
Household income-sq.	3.330E-06	1.14	8.650E-07	0.12	3.880E-06	1.20
Regions	Yes		Yes		Yes	
Years	Yes		Yes		Yes	
Constants	39.334***	10.03	46.182***	5.25	38.295***	8.31
Observations	3179		711		2468	
adj.-R2	0.063		0.085		0.030	

Note *, **, *** Statistical significant levels are 10, 5, 1%
Source calculated based on 2012–2014 Japanese Maternal Life and Work (JMLW) survey data

Table 7 Results of the influence of time poverty on happiness

	(1) Total		(2) Single mother		(3) Married mother	
	coeff.	z-value	coeff.	z-value	coeff.	z-value
Imputed time poverty	−2.782***	−5.51	−3.091***	−3.31	−2.580***	−4.14
Single mother	−1.107***	−6.24				
Age	0.258***	4.14	0.172	1.14	0.243***	3.51
Age squared	−0.003***	−4.57	−0.002	−1.06	−0.003***	−4.06
Education (ref.: Junior high school)						
Senior high school	−0.426	−1.50	−1.163**	−2.24	−0.131	−0.37
Vocational school	0.290	1.29	−0.310	−0.72	0.486*	1.76
College	0.550**	2.27	−0.151	−0.33	0.774***	2.60
University	0.089	0.31	−0.752	−1.33	0.366	1.06
Work status (ref.: non-work)						
Regular worker	1.557***	4.68	2.712***	4.19	1.204***	2.99
Irregular worker	0.854***	3.76	1.506***	3.35	0.741***	2.66
Self-employed	1.907***	5.01	3.263***	4.02	1.614***	3.53
Child number	0.451***	5.14	0.432**	2.38	0.465***	4.47
Youngest child age (ref.: aged 0–3)						
Aged 4–6	−1.007***	−5.34	−0.664	−1.23	−0.942***	−4.42
Aged more than 7	−1.752***	−6.40	−1.389**	−2.20	−1.623***	−4.97
Living with parents	0.160*	1.60	0.594***	2.79	−0.014	−0.12
Husband housework support	0.256***	9.18	−0.020	−0.22	0.291***	10.07
Household income	0.005***	5.67	0.006***	2.72	0.004***	4.48
Household income-sq.	−3.290E-06**	−2.28	−5.290E-06	−1.60	−2.110E-06	−1.32
Regions	Yes		Yes		Yes	
Years	Yes		Yes		Yes	
Constants	−4.097**	−2.12	−5.310	−1.27	−3.431	−1.49
Observations	3179		711		2468	
adj.-R2	0.115		0.047		0.107	

Note *, **, *** Statistical significant levels are 10, 5, 1%
Source calculated based on 2012–2014 Japanese Maternal Life and Work (JMLW) survey data

4.3 Comparison Between the Single Mother Group and the Married Mother Group

To consider the third issue, the difference between the single mother group and the married mother group, first, the results in Tables 5, 6, and 7 also indicate that when other factors are consistent, the single mother group is more likely to have lower well-being than the married mother group (the coefficients of the single mother dummy variable is −3.343 for subjective health status, 3.007 for mental health disorder, and −1,107 for happiness).

Second, to compare the time constraint and income effect on the probability of experiencing time poverty, (1) the coefficient of work status and youngest child age dummy variables are not statistically significant for the single women group, whereas they are statistically significant for the married women group. It is shown that the time constraint on the probability of experiencing time poverty seems smaller for the single mother group than for the married mother group. This may be because although the support from the husband is less for the single mother group than for the married mother group, the single mother group may get more support from parents and other family members. For example, based on the results in Table 1, the proportion of women who are living with parents is 32.9% for the single mother group, whereas it is 20.7% for the married mother group. It is thought when a woman with children lives with her parents, she can gain more childcare and other housework support from her parents. (2) The coefficients of household income dummy variables are not significant statistically for both the single mother group and the married women group. This indicates that the difference of income effect on time poverty is quite small between these two groups.

Third, some important findings can be deduced from comparison of the influence of time poverty on maternal wellbeing between the single mother group and the married mother group, based on the results of these two groups in Tables 5, 6 and 7. (1) For the influence of time poverty on subjective health status and mental health disorder, even though the time poverty negatively influences subjective health status and mental health disorder for both the single mother group and the married mother group, the influence is smaller for the single mother group. For the influence of time poverty on happiness, even though time poverty negatively influences happiness for both the single mother group and the married mother group, the influence is greater for the single mother group.

5 Conclusions

Using the Japanese Maternal Life and Work (JMLW) Survey data conducted in 2012 and 2014 (JMLW2012 and 2014), this study explores the determinants of time poverty and its impact on the wellbeing of the single mother and the married mother groups in Japan. The main findings for the determinants of time poverty are as follows: first, for the determinants of time poverty, the probability of experiencing time poverty is greater for the longer housework hours and the working hours groups, and the time constraint hypothesis (H1) is supported. However, housework support from other family member (e.g. husband, parents) is not statistically a significant influence on time poverty. Second, the influence of household income on the probability of experiencing time poverty is not statistically significant, therefore the income effect hypothesis (H2) is not supported. Third, the hypothesis that time poverty negatively affects maternal wellbeing (H3) is supported. Fourth, the influence of housework hours on the probability of experiencing time poverty is smaller for the single mother group than for the married mother group. The negative effect of time poverty on subjective health status and mental health disorder is smaller for the single mother group than for the married mother group. Lastly, the negative effect of time poverty on happiness is greater for the single mother group than for the married mother group.

The policy implications derived from these results are as follows. First, the results indicate that the longer working hours and time spent on housework may harm maternal health and happiness in Japan. Policies to reduce time poverty should be considered in order to increase wellbeing in Japan.

Second, results for the hypotheses tests indicate that the relation of market work, housework and discretionary activity is a trade-off, and income may not reduce time poverty even for a high-income household. This may be because the price of childcare and housework in the market is so high that they cannot buy these services. Thus, the policy to promote the support for childcare and housework from other family members and public childcare support should be considered. It is thought that the husband's participation in childcare and housework should reduce maternal time poverty. However, the effect of husband participation in childcare on time poverty is not statistically significant (see Table 4). It might be because the hours of husband participation in childcare and other housework are so few. For example, based on the results in Table 3, for the married mother group, the proportion whose husband's housework hours is 40% of the total housework hours is only 8.6%, the proportion whose husband's housework hours is more than 50% of the total housework hours is only 9.6%. The hours of husband participation in childcare and housework are fewer in Japan than in other developed countries, like Germany or France (Ma 2007).

Why are husband housework hours so low in Japan? Two reasons may explain it. First, the influence of the concept of "men for work, women for family" is widely held in Japan, thus gender role segregation exists within the household. Second, gender role segregation persists in the workplace. Most female workers are part-time workers, whereas most male workers are in full-time jobs and need to work longer hours. Thus even though a husband wants to give more support for childcare and housework, the long working hours may reduce the husband's housework hours.

Therefore, in order to address the time poverty problem for the single mother and married mother groups in Japan, it is important to promote the implementation of the Childcare Leave System for both men and women workers. Policies to mitigate the problem of long working hours should be considered. The total reform of the working hours management system, employment system, assessment system and the wage system might be necessary in the future. Policy to promote the supply of high quality and low priced childcare services in the market should also be considered to support familial wellbeing and economic growth. Although it is thought that childcare and housework support from other family members other than the husband can reduce the time constraint problem for the maternal group, the results in the study show that the effect is not statistically significant. Further research on the effects of childcare and housework support by other family members in Japan is recommended.

Finally, the results indicate that the single mother group is more likely to experience lower wellbeing than the married mother group, and the negative effects of time poverty on happiness are greater for the single mother group than for the married group. In Japan, policy for the single mother group focuses on the income poverty problem (such as the policy to promote the single mother's labor participation, and the poverty subsidy policy for the single mother group). To promote the individual's equality of wellbeing it is not enough just to implement policies that focus on the reduction of income poverty for the single mother group. Time poverty is a neglected factor. The Japanese government may productively consider policy enactment from a multidimensional poverty perspective.

Acknowledgements This research was supported by JSPS (Japan Society for the Promotion Science) KAKENHI (Grant Number JP16K03611). I am grateful to Japan Institute for Labor Policy and Training (JILPT) to provide the Japanese Maternal Life and Work (JMLW) Survey data. I also acknowledge professor Ming-Chang Tsai (Research Center for Humanities and Social Sciences Academia Sinica Taiwan) professor Yanfei Zhou (JILPT professor Akiko Oishi (Chiba University) the anonymous referees and editors for their many helpful comments and suggestions and Dr. Richard Lingwood for his helpful editing work.

References

Allen, T. D., Herst, D. E. L., Bruck, C. S., & Sutton, M. (2000). Consequences associated with work-to-family conflict: A review and agenda for future research. *Journal of Occupational Health Psychology, 5,* 278–308.

Bardasi, E., & Wodon, Q. (2010). A working long hours and having no choice: Time poverty in Guinea. *Feminist Economics, 16*(3), 45–78.

Becker, G. S. (1965). A theory of the allocation of time. *The Economic Journal, 75*(299), 493–517.

Deutsch, J., & Siber, J. (2005). Measuring multidimensional poverty: An empirical comparison of various approaches. *Review of Income and Wealth, 51,* 145–174.

Ford, M. T., Heinen, B. A., & Langkamer, K. L. (2007). Work and family satisfaction and conflict: A meta-analysis of cross-domain relations. *Journal of Applied Psychology, 92,* 57–80.

Gronau, R. (1977). Leisure, home production, and work: The theory of the allocation of time revisited. *Journal of Political Economy, 85*(6), 1099–1123.

Grzywacz, J. G., & Brenda, L. B. (2003). Work, family, and mental health: Testing different models of work-family fit. *Journal of Marriage and Family, 65,* 248–262.

Ishii, K., & Urakawa, K. (2014). Time poverty and adjusted income poverty rates by housework time in Japan. *Mita Business Review, 57*(4), 97–121 (in Japanese).

Kakwani, N., & Siber, J. (2008). *Quantitative approaches to multidimensional poverty measurement.* New York: Palgrave Macmillan.

Kalenkoski, C. M., Hamrick, K. S., & Andrews, M. (2011). Time poverty thresholds and rates for the US population. *Social Indicators Research, 104*(1), 129–155.

Liff, S. (1981). Mental health of women factory workers. *Journal of Occupational Behavior, 2*(2), 139–146.
Ma, X. (2007). Household lifetime and life inequality. In Y. Higuqi & M. Seko (Eds.), *Dynamism of household behavior in Japan III: Status, cause and its influence* (193–222). Tokyo: Keio University Press (in Japanese).
Ma, X. (2009). Does longer working hours cause mental health problem? In Y. Higuqi, M. Seko, & H. Teruyama (Eds.), *Dynamism of household behavior in Japan V: The high quality of labor market and labor participation behavior* (pp. 104–145). Tokyo: Keio University Press (in Japanese).
Matsuda, S. (2001). Gender role segregation and new gender role segregation: Double burdens of work and life. *Philosophy, 106*, 1–57 (in Japanese).
Merz, J., & Rathjen, T. (2014). Time and income poverty: An interdependent multidimensional poverty approach with German time use diary data. *Review of Income and Wealth, 60*(3), 450–479.
Perrucci, R., MacDermid, S., King, E., Tang, C., Brimeyer, T., Ramadoss, K., et al. (2007). The significance of shift work: Current status and future directions. *Journal of Family Economic Issue, 28*, 600–617.
Shockley, K. M., & Singla, N. (2011). Reconsidering work-family interactions and satisfaction: A meta-analysis. *Journal of Management, 37*(3), 861–886.
Tsuya, N. (2000). Gender differences in employment and housework: A comparative analysis of Japan, South Korea, and the United States. *Journal of Population Problems, 56*(2), 25–48 (in Japanese).
Tuttle, R., & Garr, M. (2009). Self-employment, work-family fit and mental health among female workers. *Journal of Family and Economic Issues, 30*(3), 282–292.

Xinxin Ma is an associate professor in Center for Far Eastern Studies (CFES), University of Toyama, Japan. She is a board member of the Japanese Association for Chinese Economy and Management Studies and the editor of *Japanese Journal of Comparative Economics* and *Journal of Chinese Economic and Management*. Her current research project focuses on intra-household bargaining, population ageing and family care in Japan and China. Her articles have appeared in *Journal of Happiness Studies, China Economic Review, Journal of Chinese Economic and Business Studies*. Her recent book is *Economic Transition and Labor Market Reform in China* (Palgrave Macmillan 2018) (xxma@eco.u-toyama.ac.jp).

Aging, Gender, and Migration

Changing Lives of the Japanese Elderly Under Uncertainty: An Analysis of Family Types and Economic Status

Hachiro Iwai

Abstract A high rate of the elderly participating in labor force and a high rate of their living together with their children have characterized the Japanese aging society. Although these features are still evident, a new pattern of living together has emerged since the mid-1990s: the number of living together with unmarried children has been increasingly frequent among the elderly. Economic conditions of the elderly and the child generation are related to the emergence of this pattern. Based on the quantitative analyses of SSM 2005 and JGSS 2010–15, this paper explores the changes in relations between family types and economic status for the elderly aged 60 or over. It is suggested that for low-income households, living together with children implies the mutual dependency between parent and child generation in an economic sense under uncertain conditions in recent Japanese society.

Keywords Japanese elderly · Living together with child generation · Income inequality · SSM · JGSS

1 Japan's Aging Society in Transition

The year of 1973, when the rapid economic growth in postwar Japan had ended, is called the "First Year of Welfare." In that year, the framework of policies for the elderly was formulated. Introducing the free medical care for the elderly and reforming the pension system, the Japanese government presented the future vision of aging society that all the elderly people could receive free medical care and a generous pension. At that time, rapid economic growth generated a large surplus in the annual revenue of the Japanese government, while the growth caused negative aspects such as environmental pollution. In the rural area, the support base of Liberal Democratic Party of Japan weakened due to population migration, and the progressives gained power in urban local governments. Under such circumstances, the poor living conditions of the elderly drew attention as a "shadow" aspect of rapid economic growth, revealing actual lives of the "vulnerable" left in the country that entered the rank of

H. Iwai (✉)
Kyoto University, Kyoto, Japan
e-mail: iwai.hachiro.2r@kyoto-u.ac.jp

M.-C. Tsai and N. Iwai (eds.), *Quality of Life in Japan*, Quality of Life in Asia 13,
https://doi.org/10.1007/978-981-13-8910-8_6

Western advanced countries. It is said that in the beginning of the 1970s "Japanese mass media and people were awakening to the problems of the elderly." Newspapers rapidly increased their coverage on topics of the elderly. It was hard to oppose to advocating for the support for the "vulnerable" and the "generous" policy that "the medical insurance covers everyone, the medical expense of the elderly is extremely small, and all the people receives pension" (Campbell 1992/1995).

In 1975, the people aged 65 or over were 7.9% of the total population in Japan. An irony is that although the policy was targeting on the smaller group of people, there was a rapid growth of the target population when economic growth was stagnated. In 2005, 20% of the Japanese population was over 65 years old. The ratio rose to 26.7% in 2015 and is estimated to be 36.1% in 2040. Financial crisis of pension system was expected to become more serious in as early as 1980s, owing to increasing population of the elderly, and several measures have been taken repeatedly, such as extending the age of retirement, the employment policy for the elderly, the reforms of pension system and medical care, and establishing long-term care facilities. However, the transition from "generosity" to "shouldering burden" has been so slow as to cause the current "pension crisis."

Along with the development of the elderly policy, the social status of the Japanese elderly has changed from the mid-1970s. The analyses of the SSM (Social Stratification and Mobility) data of 1975, 1985 and 1995 have already clarified the following points in regards to the elderly aged 60s or over (Iwai 1998, 2001):

First, although the ratio of labor force participation aged 60 or over is still high, the ratio of retired people in the age group increased in 1980s. In 1990s, those who worked for large companies or public offices in their 50s had a tendency to retire at the age of 60 and stop working, while the other people were likely to continue to work.

Second, the improvement of public pension system enhanced the economic status for retired people. In 1975, the majority of the retired elderly people had low income, but in 1985, those people were divided into a low-income group and a group of pensioners. In 1995, the income distribution of retired people was centered on the amount of pension. It is the 1990s when the retired elderly broke away from economic dependency on others.

Third, the meanings of family types for the elderly have changed. There were a significant improvement of economic status of retired people and a subsequent reduction of economic dependency on other members in their household. As a result, the number of people living together with child generation as a form of economic dependency decreased sharply. In other words, the modern conjugal family became one of choices that the retired elderly could make.

The early 1990s might be the "golden age" for Japanese elderly people. However, after the late 1990s, economic recession has deprived the elderly of employment opportunities, and their income level has not improved. Dwindling birthrate and aging population force us to revise the pension system as well as the medical and nursing care systems. The image of healthy, independent and rich elderly people seems to be "declining." In a comparative perspective, a high employment rate among the elderly and a high proportion of the elderly living together with their children have characterized the Japan's aging society. However, the proportion of the latter has been declining. Currently two opposite trends are evident: among persons aged 60 or over, the proportion of living with the child's couple is declining, while that of living with unmarried children is rising.

This paper aims to clarify the trends of economic conditions among the Japanese elderly, analyzing both SSM 2005 and JGSS 2010, 2012 and 2015. It examines how the social status of the elderly with relation to their family types has changed after the late 1990s. The results of analyses show that the average income and household income among the older men decline from 1995 to 2010–15. Then, seeing the trends of Gini coefficients reveals that relatively high levels of income inequality exist among men aged 60–64, and that after the age, the level of inequality is declining. In addition, the analyses of SSM 2005 find that there are three types of retired men living with children. The first type is the elderly receiving economic support by living together with children. The second one is the elderly supported by their own pension, living with low-income children. The third one is the affluent elderly supported by their own pension, living with better-income children. Three types of the retired elderly are also identified in JGSS 2010–15, although the second type was not found in the 1995 data. Based on these results, this paper argues that economic uncertainty facing young people in the 2000s changes the meaning of the elderly living together with their children. It is argued that the intergenerational mutual dependency through living together is important for low-income households as a way of life, which enables them to withstand uncertain conditions.

Before introducing specific results of the analyses, it would be helpful to take a glance at the backgrounds of this study, i.e., characteristics of the Japan's aging society in a comparative perspective.

2 Welfare Regimes and Japan's Family Changes

Characteristics of Japanese aging society, compared with other countries, are the rapid pace of aging, the higher ratio of workforce participation and the high ratio of the elderly living together with their children. In recent years, on the other hand, there are obvious changes in both labor participation and living together with children.

Figure 1 shows changing patterns of labor force participation of men aged 60–64 in Sweden, Germany, the US and Japan. Differences among four countries are not clear in 1970; labor force participation rates of four counties ranged from 70 to 80%. However, the feature of each country had been evident in declining patterns of the rate from the 1970s to the 2000s. In each country, the participation rate was declining, and the decline was most prominent in (former West) Germany. The rate was also declining in the US and Sweden, but not as sharp as in Germany, and it remained

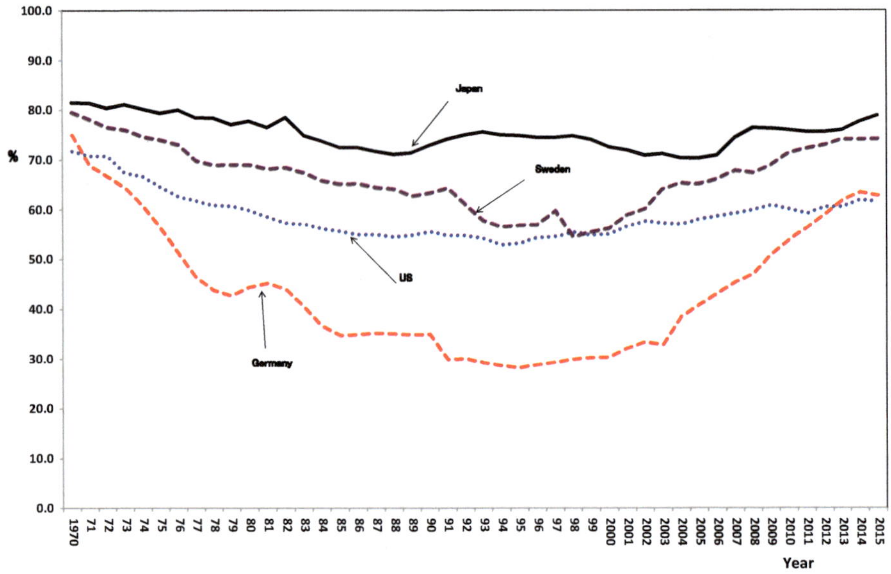

Fig. 1 Comparison of labor force participation rate among men at age 60–64: 1970–2015. *Source* OECD Labor Force Statistics (OECD 2005, 2019)

higher in Sweden than in the US. In addition, the rates had declined until the 1990s in all the three countries and tend to swing back thereafter.[1]

Japanese men show the highest participation rate in the four countries. This is because the rate in Japan since 1970 has been most stable; there was a slight decline from 1970 to the mid-1980s, but the decline stopped in the 1990s. In recent years, the rate remains unchanged and then is gradually rising. It was 70.3% in 2005 and 78.9% in 2015. From a comparative perspective, the Japanese society from 1970 to 2010 was very unique in terms of the highest rate of labor force participation for the elderly. However, the uniqueness has been recently vanishing since the rates of working elderly people are rising in other countries.

The three countries shown in Fig. 1, Germany, the US and Sweden, are representing three types of "Welfare Regimes" proposed by Esping-Andersen.[2] Differences among three countries are a result of policy responses to employment situations and welfare conditions after the Oil Crisis in 1970s.

In Sweden, categorized as the "Social Democracy", the welfare state provides a wide range of services. However, ensuring the personnel and financial basis for welfare requires the level of employment as high as possible. Because both men and women tend to remain in labor market until they reach the retirement age and the policies do not encourage early retirement, the labor force participation rate of people

[1]See Kohli et al. (1991) for comparative research of early retirement. For comparative research using individual data, see Blossfeld et al. (2006).

[2]The following description relys on Esping-Andersen (1999/2000).

in their early 60s is ascending. The participation rate of the elderly is also high in the "Liberalist" US, but here the retirement would be an outcome of contract between a person and a company, and the market mechanism tends to determine the individual pension and the quality and quantity of welfare services received. Therefore, there is not a standardized pattern of retirement. The participation rate is the lowest in Germany, a representative of the "Conservatism". Here the career of breadwinners (men) comes first, and their benefits are expected to cover the welfare of all family members, depending on types and fields of occupation and social insurance. There were few policies expanding employment of women and the elderly. After the mid-1970s, early retirement of the middle-aged and elderly was encouraged in order to assure employment opportunities for young people, and the pension system for the elderly was well arranged. The system was called "the welfare state without labor" for the elderly, and the participation rate of people aged 60 or older was markedly low. Recently, however, this system has faced a turning point.

Japan did not follow the directions that Sweden, Germany and the US took, so that it has distinct features. Although the tendency of early retirement was recognized, the labor force participation rate has been consistently high for the 30 years. In large companies or public offices, the age of retirement was 60 years old in general, but the age tends to rise. Self-employed workers do not have to retire, and many people in large companies or public offices continue to work by being reemployed or moving to small-medium enterprises. In addition, there are measurements promoting employment of the elderly.[3] These provided the backgrounds for the uniqueness of Japan's aging society (Ebbinghaus 2001, 2006).

Esping-Andersen categorized Japan as a family-oriented welfare state, based on the persistence of gender division of labor, the higher rate of the elderly living together with child generation, and the deep-rooted belief that the welfare of people was generated from family ties (Esping-Andersen 1987). "Family-oriented" here does not mean promoting formation of family and enhancing birthrate, but reproducing mutual dependency among family members. Under the family-oriented welfare regime, the expansion of women's opportunities is not consistent with the family formation. Women are forced to choose either remaining in home as a housewife to take care of children and the elderly, or continuing their job and thus postponing family formation. Therefore, when the service economy expands, the educational level of women improves and their years at work extend, the cost of exit from work life becomes larger, so that marriage tends to be postponed and birthrate continues to decline. Not only Japan but also South European Catholic countries, such as Italy and Spain are classified into the "family oriented" countries, although Japan and South European countries are different in labor force participation rate of the elderly.

I would like to call the Japanese life pattern revealed from the cross-national comparison "the post-war Japanese life course." The elderly had been characterized by the life patterns of continuing to work and living together with their children. In Japanese society, adult men were assured stable job and income until they became

---

[3]Seike and Yamada (2004) is a research based on their interest on promoting employment of the elderly.

aged, and the degree of dependence on pension system was low as long as family members offered care for children and the elderly. Thus, the family-oriented welfare regime was sustained, and the post-war Japanese life course had been stable from 1970s to the 1990s. However, social changes after the late 1990s, such as realignment of Japanese employment system, expanding employment opportunity for women, the decline of birthrate and population aging, force to change the conventional patterns of life.[4]

The higher ratio of living together with their children among the Japanese elderly is still observable from a cross-national comparison, but the ratio has been consistently declining and the new trends have been emerging. Figure 2 shows the trends in distribution of persons aged 65 years and above by types of family from 1980 to 2015.

Four characteristics of family changes should be identified. First, the ratio of living with child's couple has been very sharply declining from 52.5% in 1980 to 12.5% in 2015. Second, the ratio of households of couple only has been rising from 19.6% in 1980 to 38.9% in 2015. Third, the ratio of living with unmarried children has been also going up from 16.5% in 1980 to 26.5% in 2015. Fourth, the ratio of one-person household has been gradually rising from 8.5% in 1980 to 18.0% in 2015.

Focusing on the trends in the distribution of households with persons aged 65 years or over also reveals that a large scale of family changes occurred for the Japanese

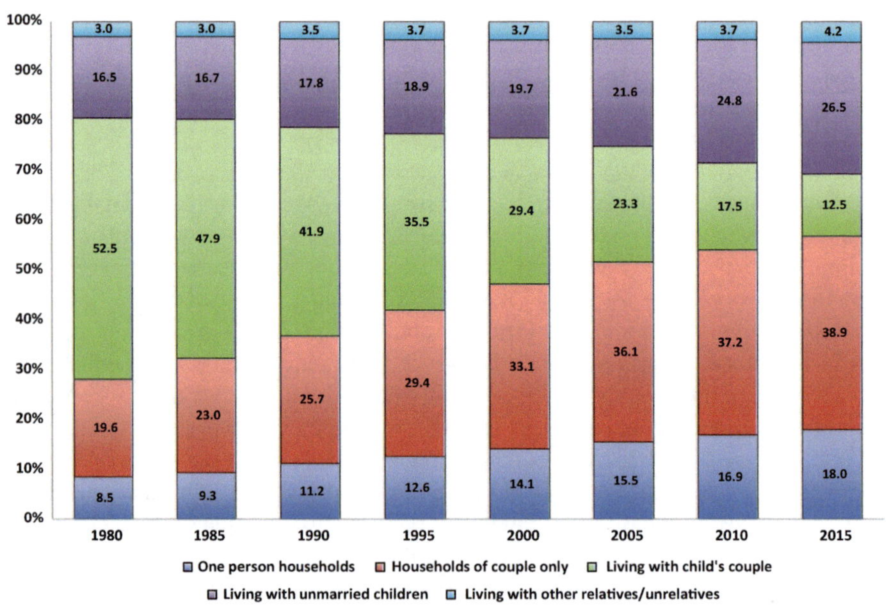

Fig. 2 Trends in percent distribution of persons aged 65 years and above by types of family: 1980–2015. *Source* Vital Statistics 2018 by National Institute of Population and Security Research

[4]As to the development of life course study, see Iwai (2006a, b).

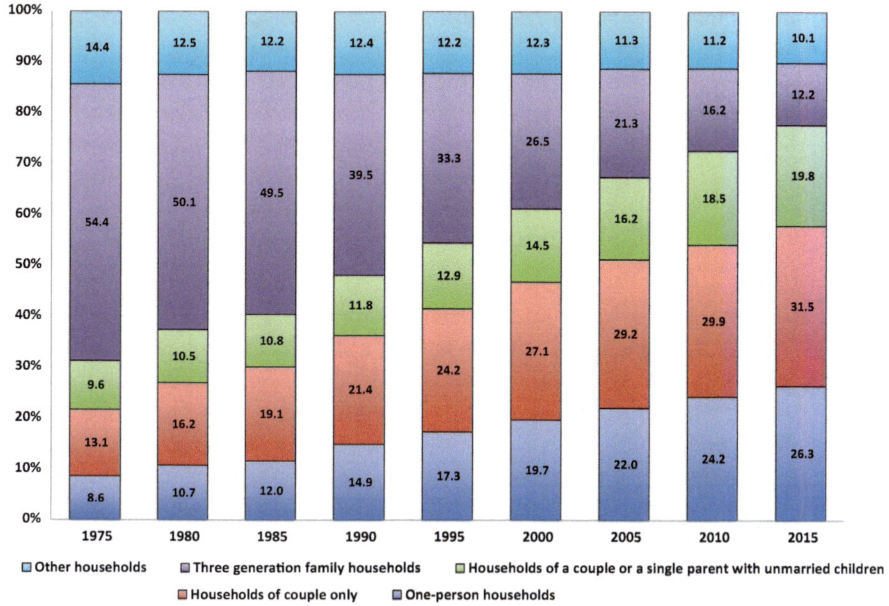

Fig. 3 Trends in percent distribution of households with persons aged 65 years and above by structure of household: 1980–2015. *Source* Vital Statistics 2018 by National Institute of Population and Security Research

elderly. Figure 3 shows that in 1980, 50.1% of households with persons aged 65 years or over were three-generation household. However, since then, the ratio has been declining to 12.2% in 2015. On the other hand, the ratio of households of a couple or a single parent with unmarried children has changed from 10.5% in 1980 to 19.8% in 2015. Three-generation household is no longer a dominant family type in Japan. The Japanese households of the elderly can be categorized into 4 types: the household of couple only, the one-person household, the household of a couple or a single parent with unmarried children and the three generation household.

3 Changes in the Economic Status of the Elderly

This section explores how family types are related to economic conditions of household members, using SSM (Social Stratification and Mobility) Surveys in 1985, 1995 and 2005,[5] and JGSS (Japanese General Social Survey) in 2010, 2012 and 2015.[6]

[5]SSM Surveys use nationally representative sample aged from 20 to 69. The 2005 survey targeted for 13,031 cases and the response ratio is 44.1%; 5742 cases are valid. Male cases are 2660.

[6]JGSS 2010, 2012 and 2015 are also based on nationally representative samples aged from 20 to 89, targeted for 9000 cases (2010), 9000 (2012) and 4500 (2015). The response rates are 62.2, 59.0

Table 1 The percentages of "not working" men aged 55 and above: SSM 1985–2005 and JGSS 2010–15

Age group	SSM 1985	SSM 1995	SSM 2005	JGSS 2010–15
55–59	10.0	6.0	8.2	10.2
60–64	30.3	28.1	32.3	29.1
65–69	42.9	39.1	46.8	50.7
70–74				67.6

Figure 1 showed that the labor force participation rate of Japanese men aged 60–64 had been consistently maintained high in spite of a slight decline in 1980s and that the rate has been gradually rising after the mid-2000s. In this section, first of all, I would like to examine whether or not this tendency is confirmed in the SSM and JGSS data. Table 1 shows the proportions of the "not working" in the employment status of the elderly respondents surveyed in the three waves of SSM, classified into three groups: aged 55–59, 60–64, and 65–69, and the proportions of aged 55–59, 60–64, 65–69, and 70–74 from the JGSS 2010–15 are added. In 1995, the proportions of all the three groups decreased from 1985, but they all rose in 2005 from 1995. JGSS 2010–15 indicates that more than half of men aged 65 or over did not work. Although policies for promoting employment of the elderly are advocated, employment opportunities of the elderly have not been expanded since the late 1990s. Furthermore, JGSS 2010–15 asked "not working" respondents whether they "are seeking for job." About 16.6% of male respondents aged 60–64 and only 3.3% of male respondents aged 65–69 answered "yes."

By the mid-1990s, the expansion of pension system improved the economic status of the not-working elderly and lowered their necessity of economic dependence for men aged 60 or over without working. However, due to the economic recession after the mid-1990s, the income level of average Japanese people has not risen. Is there any change in the economic status of people in their 60s or not?

SSM and JGSS ask respondents about the incomes of respondents themselves, their spouse and their household. Although there is no information on specific sources of income, the surveys asked respondents to answer the amount of income from work and the income from temporary or additional earnings. In regard to the elderly, military pension, regular pension and retirement benefit are included in the amount.[7]

Table 2 shows respondents' own and household income (annual) in SSM 1985, 1995, 2005 and JGSS 2010–15.[8] Respondents are divided into three age groups:

and 52.6%. This paper uses an integrated data set of 2010, 2012 and 2015 that use the same format of collecting the information about income and household composition.

[7]Income is surveyed not by the actual amount but by the income code. For example, the income code indicating more than 1.5 million yen and less than 2.5 million yen is calculated as 2 million yen. Income code varies among waves of the survey.

[8]All figures are adjusted by the 2015 consumer price as 100. The indexes are 85.4 in 1985, 97.6 in 1995, 96.9 in 2005, 96.5 in 2010, 96.2 in 2012 and 100 in 2015. In case of JGSS 2010–15, after actual incomes of each survey data are adjusted, average incomes are calculated.

Table 2 Average own income and household income of Japanese men aged: 55–74: SSM 1985, 1995, 2005 and JGSS 2010–15

Age		SSM 1985	SSM 1995	SSM 2005	JGSS 2010–15
55–59	Own income	470.20 (401.55)	**675.56 (659.35)**	576.45 (558.58)	*565.95 (548.21)*
	Household income	711.65 (607.75)	**962.97 (939.86)**	748.71 (725.50)	*732.46 (709.37)*
60–64	Own income	355.80 (303.85)	**478.57 (467.09)**	451.60 (437.60)	*417.24 (404.24)*
	Household income	572.84 (489.20)	**748.20 (730.24)**	612.13 (593.15)	*571.52 (553.69)*
65–69	Own income	333.07 (284.44)	**416.27 (406.28)**	334.25 (323.89)	*325.72 (316.72)*
	Household income	533.03 (455.21)	**697.62 (680.88)**	461.44 (447.14)	*463.93 (451.19)*
70–74	Own income	No data	No data	No data	*313.58 (304.32)*
	Household income				*442.00 (429.22)*

Notes
1. Ten thousand yen (1$ ≒ 110 yen)
2. Figures are adjusted by the 2015 Consumer Price Index
3. Figures in parentheses are not adjusted

aged 55–59, 60–64 and 65–69. JGSS 2010–15 has the data for men aged 70–74. Both working and not-working respondents are included in all the groups. This table indicates the considerable decreases of both individual and household income from 1995 to 2005 in all the age groups after both increased sharply from 1985 to 1995. Especially, it is clear that the individual income of the respondents in their late 50s decreased one million yen from 6.75 million to 5.76 million yen, and that the household income declined by more than two million yen. The income levels aged 60 or over do not improve in JGSS 2010–15, rather declining.

Next, Table 3 examines the trends of income inequality by Gini coefficients in each age group from 1985 to 2010–15. As for the own income, the levels of inequality in three age groups are highest in 1985. Then the levels declined in 1995 and did not change in 2005, excluding very-high-income earners. The Gini coefficient of the own income in age 60–64 seems higher in 2010–15.

Improvement of the annual income from 1985 to 1995 narrowed the income disparity among the Japanese elderly. In 2005 and 2010–15, the Gini coefficients of men aged 60–64 slightly rose. At the age of 60, Japanese male workers employed by large companies or public sectors are forced to transit from work to retirement or find another employer. Some could keep earning a good income. Others would face declining income till the pension age at 65. It is thought that the rise of income disparity in the early 60s reflects this situation in 2010–15.

Table 3 Income inequality by Gini coefficient

Age		SSM 1985	SSM 1995	SSM 2005		JGSS 2010–15
55–59	Own income	0.409	0.372	0.350		0.368
	Household income	0.325	0.337	0.313		0.340
60–64	Own income	0.458	0.395	**0.419**	**0.389**[a]	**0.418**
	Household income	0.369	0.393	0.398	**0.390**[a]	0.374
65–69	Own income	0.516	0.416	0.357		0.346
	Household income	0.381	0.381	0.357		0.331
70–74	Own income	No data	No data	No data		0.389
	Household income	No data	No data	No data		0.349

[a]Without more than 35 million yen

Gini coefficients of the own income among old men were much higher than those of the household income in 1985. At that time, the Japanese elderly tended to live together with child's couple, so that other members of household could financially support them. High level of income disparity was not a serious social problem. After 1995, the differences of Gini coefficients between the own income and the household income became smaller due to the decline in the ratio of living together with child's couple.

Table 4 shows the trends of Gini coefficients for the not-working respondents and the working respondents respectively at the early 60s and at the late 60s. For the working men, Gini coefficients of the own income at the age 60–64 and at the age 65–69 had declined from 1985 to 2005, but it did not change very much in 2010–15. On the other hand, for the not-working men, the Gini coefficient at the age 60–64 rose from 1995 to 2005, while the value at the age 65–69 declined from 1985 to 2010–15. The declining income disparity among the not-working men aged 65 or over means that the improvement of pension system tends to standardize their economic status. However, the high level of income disparity among the not-working men at the age 60–64 suggests that some received good pension and others not. The Gini coefficients of household income for the not-working men at the age 60–64 are also relatively high in 2005 and 2010–15.

In 1985, the Gini coefficients of the own income were much higher than those of the household income among the not-working men. This evidence also suggests that lives of not-working old men and women tended to be financially helped by other members of household, i.e. child's couple. However, the differences of Gini coefficients between the own and the household income are not clear after 1995. After the 2000s, more and more older women keep working at the age of 60s. Job opportunities for the elderly women would influence financial conditions of household.

Table 4 Gini coefficients by working and not-working

Age			SSM1985	SSM1995	SSM2005		JGSS 2010–15
60–64	Not working	Own income	0.461	0.354	**0.498**	**0.456**[a]	**0.450**
		Household income	0.315	0.359	**0.488**	**0.453**[a]	**0.407**
	Working	Own income	0.421	0.377	0.363		0.390
		Household income	0.370	0.375	0.341		0.355
65–69	Not working	Own income	0.501	0.353	0.289		0.270
		Household income	0.365	0.369	0.334		0.283
	Working	Own income	0.481	0.400	0.379		0.380
		Household income	0.347	0.355	0.361		0.342

[a]Without more than 35 million yen

4 Family Types and Economic Status for the Not-working Elderly

In case that the Gini coefficient of the household income is lower than that of the own income among the not-working old people, economic conditions of their life could be helped by other members of household living together. This type of economic intergenerational relation was found in 1985. However, due to expanding pension system, less and less older people financially depend on their children. According to the results of SSM 1995, 50.3% among not-working men aged 60–69 lived together with their child. However, the ratio of SSM 2005 declined to 37.2% and in JGSS 2010–15 the rate is 33.1%. In addition, the economic conditions of younger generations have not been improved after the mid-1990s. As a result, the meaning of the elderly living with children should be reconsidered from a different perspective.

Figure 4, which is based on SSM 2005, shows the annual income distribution of the not-working men aged 60–69 living with their children and the distribution of their household's annual income. In addition, for the purpose of comparison, Fig. 5 shows the same data a decade earlier, based on SSM 1995.

Figure 4 indicates that there are three types of the not-working elderly living with their children. First, about 20% are the low-income elderly whose own annual income is 1.2 million yen or less, named as Type 1. Since the proportion of the elderly whose household income falls in the same range is lower, this type as a whole receives economic support by living with their children.

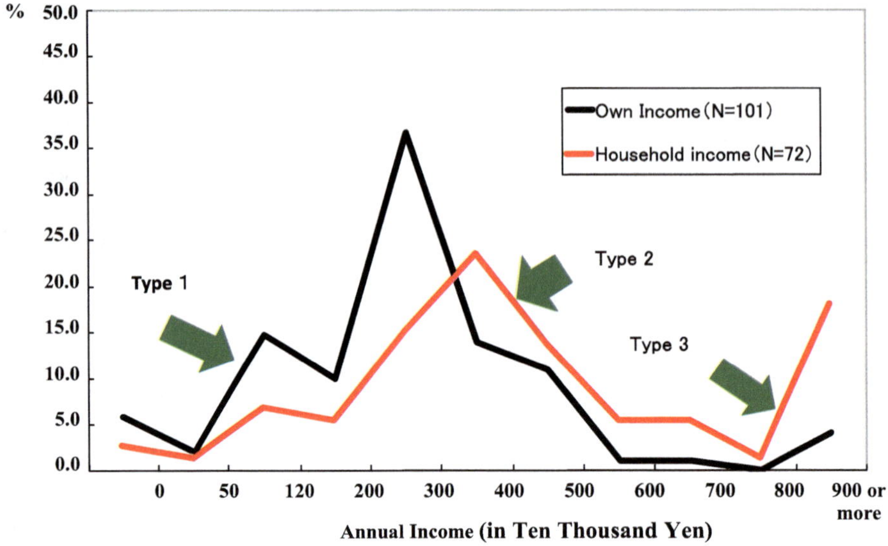

Fig. 4 Income distribution of not-working men aged 60–69 living with their children in 2005 based on SSM 2005

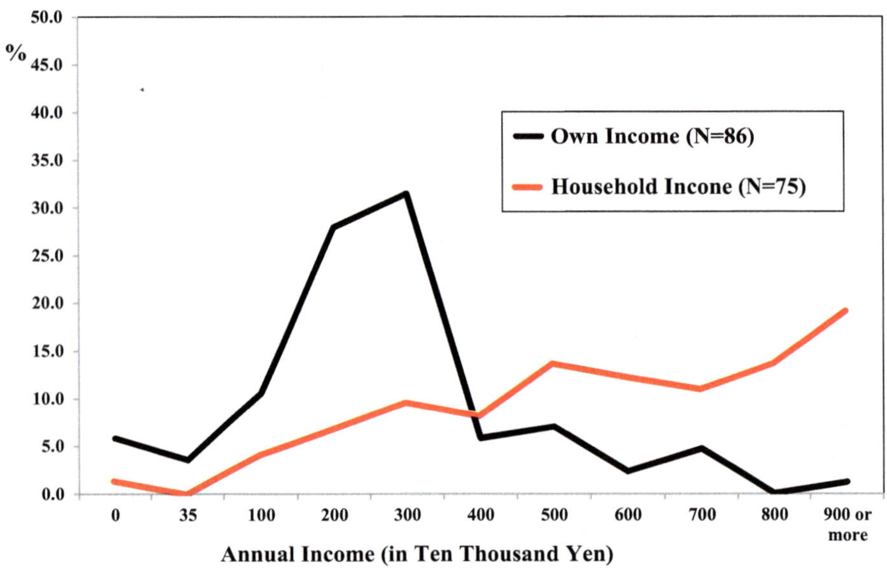

Fig. 5 Income distribution of not-working men aged 60–69 living with their children in 1995 based on SSM 1995

Second, more than 20% of the elderly concentrate around the household income of 4.0 million yen (Type 2). Their own income is concentrated in the range of 2.0–3.0 million yen, that is, around the amount of their pension. Therefore, despite living with their children, their household income is not significantly increased. This suggests that the income of children is not high. Third, approximately 18% of the elderly have a household income of 9.0 million yen or more. They clearly benefit from living with their children (Type 3).

Comparing these three types with the result based on the 1995 data shown in Fig. 5 reveals the changes occurred among the not-working elders.[9] The proportion of the low-income elders is a little higher in 2005, but the own annual income of the elderly is concentrated in the range of 2.0–3.0 million yen. There is no substantial difference between the 1995 and 2005, own income distributions. As for the household income, about 60% had 5.0 million or more yen in 1995, but the proportion was lower in 2005 at around 30%. Close to 20% had an annual household income of 9.0 million yen or more in both 1995 and 2005, showing little change. The tendency that the not-working men in their sixties living with their children would rely economically on them was very weak in 1995. In comparison with the case of 1995, it can be considered that in the case of 2005 a mutual economic reliance between parents and children existed behind the scenes of the not-working elderly living with their children. It implies that children could maintain economic stability by living with their parents rather than living separately from them.

The second type of the elderly in 2005 was not identified in the data from the 1995 survey. In 1995 it seemed that households with elderly members and their children lived relatively comfortably, having the children's income plus the parents' pension. In the case of 2005, a new style of living emerges in which parents, who rely on pension income, and low-income children become mutually dependent by living together. Under the traditional family-oriented system, the children's generation is supposed to support the elderly. However, the uncertainties facing young people since the second half of the 1990s have changed the meaning of the elderly living with their children. At present, the intergenerational mutual dependency through living together is important for low-income households as a way of life, which enables them to withstand uncertain conditions.[10]

Using JGSS 2010–15, Fig. 6 shows the results from the same analysis: the own annual income distribution of the not-working men aged 60–74 living with their children and the distribution of their household's annual income. The characteristics of the not-working elderly living with children in 2010–15 are almost identical with those in Fig. 5, although some differences are found. Their own income is concentrated in the range of 2.0–3.0 million yen, higher than in Fig. 4, while two small peaks of household income are found in the ranges of 2–3 million yen and 4–5 million yen.

Suppose that the annual income of child living together is around 2–3 million yen, and that the pension of the elderly is around 2–3 million yen a year. Combining the pension of the not-working parent and the child's income would make the household

---

[9]Income codes are slightly different between the 1995 and the 2005 surveys.

[10]See Iwai (2011) for analysis of the social status and income gaps of the elderly.

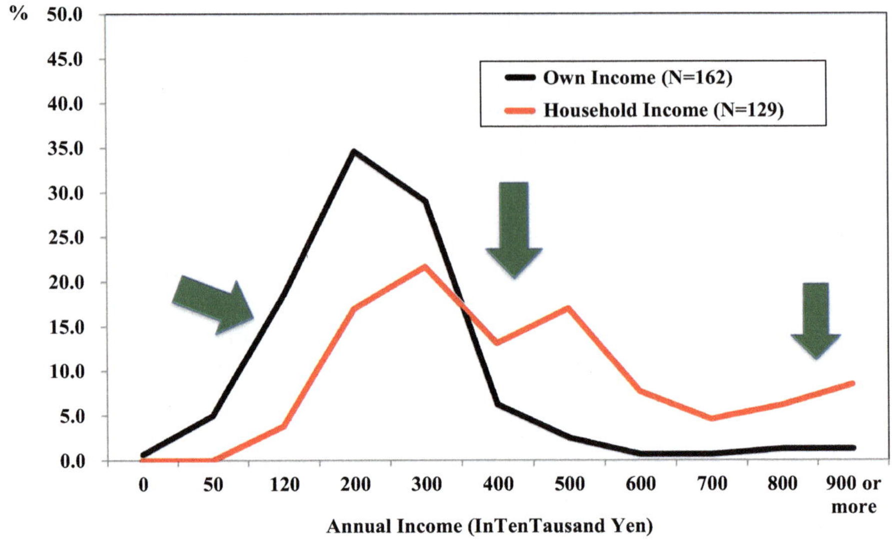

Fig. 6 Income distribution of not-working men aged 60–74 living with their children in 2010–15: JGSS 2010–15

income three to five million yen. In Fig. 6, the annual income of child living together is not high. The ratio of Type 3, the elderly having household income of higher than 9 million yen, is 8.5%, which is lower than in Fig. 5. The rich elderly that economically benefit from living together with their children is no longer typical.

While a decent economic life of the not-working elderly has been maintained by the pension system, the economic conditions of the younger generation have not improved in the early 2010s. The intergenerational mutual dependency through living together has been adopted for low-income households as a way of life.

5 Conclusion

A high rate of the elderly participating in labor force and a high rate of their living together with their children have characterized the Japanese aging society. Although these features are still evident, a new pattern of living together has emerged since the mid-1990s: the number of living together with unmarried children has been increasingly frequent among the elderly. Economic conditions of the not-working elderly and the child generation are related to the emergence of this pattern. Based on the quantitative analyses of SSM 2005 and JGSS 2010–15, this paper explored the changes in relations between family types and economic status for the elderly aged 60 or over.

The main findings are the following: first, individual and household incomes of the elderly had declined sharply from 1995 to 2005. While expanding pension system keeps the economic status of the not-working elderly stable, the high level of income disparity among the not-working men aged 60–64 is evident in the 2000s and the early 2010s. The early 60s of age for not working men is now an unstable transition time in that some receive good pension and others are likely to struggle to survive till the pension age.

Second, the rate of living together with children among the elderly is declining and the not-working elderly are more likely to live separately from their children. Expanding pension system ensures the economic stability, and there is no change in this tendency. However, examining income distributions of the not-working elderly living with their children clarified the new pattern of living together emerged from 1995 to 2005. The distribution of household income showed that living together with children in 1995 could be an affluent living style. However, the household income did not increase in 2005 due to the stagnated income of child generation. As a result, the meaning of living together with children has changed. For low-income households, living together with children implies the mutual dependency between parent and child generation in an economic sense. This pattern is confirmed in the early 2010s.

According to the typology of Esping-Andersen, Japan is classified as the family-oriented welfare regime. The results of analysis in this paper indicate that reproducing mutual dependency among family members remains in low-income families of living together with children. However, it is important to point out that this life pattern has emerged after the late 1990s, due to the economic instability of child generation.

By the early 1990s, the economic status of the elderly had rapidly improved in Japan. One could choose whether to continue work life or not, and whether to live together with one's child or not at the later stage of life. The image that the elderly could freely seek new objectives of life had been widespread and the opportunities of making it possible had expanded. However, from the late 1990s this image has been forced to recede in terms of uncertain economic conditions. As the birth rate declines and the population ages rapidly, it is predicted that the economic burdens of maintaining the welfare system increase furthermore. Under such circumstances, it is necessary to establish a self-defensive lifestyle (i.e. the maintenance of economic stability, keeping healthy life and keeping intimate human relationships) between generations.[11]

Acknowledgements 1. The 2005 SSM Survey Committee permits the use of data from the Social Stratification and Mobility (SSM) Survey.
2. The Japanese General Social Survey (JGSS) are designed and carried out by the JGSS Research Center at Osaka University of Commerce (Joint Usage/Research Center for Japanese General Surveys accredited by Minister of Education, Culture, Sports, Science and Technology) in collaboration with the Institute of Social Science at the University of Tokyo.

[11]See Iwai (2013) for life course changes after the mid-1990s.

References

Blossfeld, H.-P., Buchholt, S., & Hofacker, D. (Eds.). (2006). *Globalization, uncertainty and late careers in society*. London: Routledge.

Campbell, J. C. (1992). *How policies change: The Japanese government and the aging society*. Princeton UP. Japanese edition in 1995, translated by H. Miura, & S. Sakata, *Nihon seifu to koreika shakai: seisaku tankan no riron to kensyo* [Japanese Government and aging society: Theory of policy change and examination]. Chuo hoki.

Ebbinghaus, B. (2001). When labour and capital collude: The political economy of early retirement in Europe, Japan and the USA. In B. Ebbinghaus, & P. Manow (Eds.), *Comparing welfare capitalism: Social policy and political economy in Europe, Japan and the USA* (pp. 76–101). London: Routledge.

Ebbinghaus, Berhard. (2006). *Reforming early retirement in Europe, Japan, and the USA*. New York: Oxford University Press.

Esping-Andersen, Gøsta. (1987). Hybrid or unique?: The Japanese welfare state between Europe and America. *Journal of European Social Policy, 7*(3), 179–189.

Esping-Andersen, G. (1999). *Social foundations of postindustrial economies*. New York: Oxford University Press. Japanese edition in 2000, translated by M. Watanabe, & K. Watanabe, *Post kogyo keizai no shakaiteki kiso – Shijo, fukushi kokka, kazoku no seiji keizaigaku* [Social foundations of postindustrial economies – Political economics of market, welfare state and family]. Sakurai shoten.

Iwai, H. (1998). Karei to life course – konenreisou no shokugyokeireki, shotoku, kazoku ni kansuru bunseki [Aging and life course – Analyses on career, income and family of the elderly]. In H. Iwai (Ed.), *1995 nen SSM chosa series 13 gender to life course* [1995 SSM survey series 13 gender and life course]. 1995 nen SSM kenkyukai (pp. 87–112).

Iwai, H. (2001). Koreisha no shakaiteki chii no tenkan [Transition of the elderly in social status]. *Riron to hoho* [*Sociological Theory and Methods*], *16*(2), 211–227.

Iwai, H. (2006a). Life course kenkyu no 20 nen to keiryo shakaigaku no kadai [20 years of life course study and issues of sociometrics]. *Riron to hoho* [*Sociological Theory and Methods*], *21*(1), 13–32.

Iwai, H. (2006b). Post-fordism jidai no life course to kyoiku. In K. Inagaki (Ed.), *Kodomo, gakko, shakai kyoiku to bunka no shakaigaku* ("Life course and education in post-fordism era" in *Children, school and society – Sociology of education and culture*) Sekaishisosha (pp. 277–301).

Iwai, H. (2011). Koreisha no shakaiteki chii to kakusa [The social status of the elderly and disparity]. In Y. Sato, & F. Ojima (Eds.), *Kakusa to Tayousei, Gendai no Kaisou shisutemu Vol. 1* [Disparity and Diversity, Contemporary Stratification System Vol. 1]. University of Tokyo Press (pp. 191–206).

Iwai, H. (2013). Shrinking of the Japanese uniqueness: A quantitative analysis of life course changes. In O. Emiko, & H. L. Aoi (Eds.), *Transformation of the intimate and the public in Asian modernity* (Chapter 4, pp. 116–143). Brill.

Kohli, M., Rein, M., Guillemard, A.-M., & van Gunsteren, H. (1991). *Time for retirement: Comparative studies of early exit from the labor force*. Cambridge University Press.

OECD. (2005). Labour force statistics 198–2004.

OECD. (2019). Labour market statistics: Labour force statistics by sex and age: indicators. *OECD employment and labour market statistics* (database).

Seike, A., & Yamada, A. (2004). *Koreisha shuro no keizaigaku* [Economics of labor participation by the elderly]. Nikkei.

Hachiro Iwai is Professor of Sociology of Education, Graduate School of Education, Kyoto University. His research areas are sociology of education, life course sociology, social stratification, and social research methods. The current research topic focuses on the effects of expanding educational opportunities on subsequent life course outcomes after the 1990s in Japan, using quantitative life history data.

Ancestor Worship and Quality of Life: Transforming Bonds with the Deceased in Contemporary Japan

Masayuki Kanai, Katsumi Shimane and Dang Thi Viet Phuong

Abstract Literature in cultural psychology on happiness has argued that the connectedness and interdependence of the self with others is a key feature of East Asian societies. Ancestor worship is regarded as the most significant social institution in East Asian societies sharing Confucian tradition, such as Japan and Vietnam. Ancestor worship is a kind of reciprocal system in which the living and the deceased exchange mutual care. It links one's self to the ever-continuing family lineage, providing a basis for an individual's identity. In contemporary Japan, such beliefs and customs are rapidly diminishing due to drastic changes in population and family structure since the Second World War. This chapter addresses the question that how people in Japan cope with the situation in which the traditional assurance for their afterlife does not work well, and identifies the ways in which they seek alternative bases for their identity by examining the case of a new trend in burial. The new trend in death preparation indicates a silent transition from "bonds with the deceased" to "bonds with the living" in contemporary Japanese society. Both types of bonds can assure one's sense of self and the meaning of life and fit the contextual features of East Asian societies.

Keywords Ancestor worship · The self · Bonds with the deceased · Bonds with the living · Shrinking family

A vast volume of empirical literature on the quality of life has shown that religion and spirituality is related to subjective wellbeing (Inglehart 2010; Marques et al. 2013; Sawatzky et al. 2005). Some cultural anthropologists argued that ancestor worship is a kind of spirituality featured in Japanese society (Plath 1980; Smith 1974). Takeda (1976) discussed ancestor worship has grounded traditional Japanese

M. Kanai (✉) · K. Shimane
Senshu University, Tokyo, Japan
e-mail: mkanai@senshu-u.jp

K. Shimane
e-mail: shimane@isc.senshu-u.ac.jp

T. V. P. Dang
Vietnam Academy of Social Sciences, Hanoi, Vietnam
e-mail: dangvietphuong@ios.org.vn

M.-C. Tsai and N. Iwai (eds.), *Quality of Life in Japan*, Quality of Life in Asia 13,
https://doi.org/10.1007/978-981-13-8910-8_7

household system, *ie*, even though it is not necessarily connected to specific religions. He argued that ancestor worship compels people to continue ancestral rites based on the supreme value that households must be preserved forever. The household system places individuals within a long-lasting imagined lineage from their ancestors to themselves. Komoto (2001) further pointed out that ancestor worship is even a spiritual basis of the self and identity of the people living in traditional Japanese society.

The concept of self is also related to a debate about cultural differences between the West and the East. Recent empirical research in cultural psychology and anthropology has found that East Asian culture accentuates interdependence and connectedness among individuals, whereas Western culture emphasizes independence of individuals (Markus and Kitayama 1991; Uchida et al. 2004). This dichotomy was also highlighted by some Japanese social scientists in 1970s, when Japan enjoyed economic prosperity (Hamaguchi 1977; Kumon 1982). Moreover, it has been synchronized with arguments on the features of "Confucian Asia" (Inoguchi and Shin 2009) including Japan, Korea, China, Taiwan, and Vietnam (Tu 1996; Yum 1988).

However, a drastic change in the population and family structure in Japan in the late twentieth century has prevented the traditional household system from functioning as before. Conducting frequent memorials for the ancestors has become more and more difficult nowadays, which lowers people's expectation that they will also be worshiped by their descendants in turn. This transformation of the practice and attitude of ancestor worship might deteriorate a basis of the self and harm subjective wellbeing of individuals in contemporary Japan.

The aim of this chapter is to see how Japanese people cope with the current situation, in which the traditional assurance for their afterlife does not work well, and to identify the ways in which they seek alternative bases for their identity: this is accomplished by examining the case of a new trend in burial.

The next section deals with our preliminary argument to confirm there certainly exists micro-level association between ancestor worship and subjective wellbeing in four Asian societies under Confucian tradition, using cross-national survey data which first-ever developed instruments for practice and attitude toward ancestor worship. The following section overviews historical contexts in Japan and Vietnam, who are in the opposite sides among the four Confucian societies examined in the previous section, focusing on the changing population and family structure. The subsequent section scrutinizes emerging practice and attitude toward funerals and burial among single people who are excluded from the traditional household system which was supported by ancestor worship. The concluding section interprets its meaning as a shift from "bond with the deceased" to "bond with the living."

1 Ancestor Worship and Subjective Wellbeing in Confucian Asia: Empirical Evidences

This section verifies a micro-level correlation between ancestor worship and subjective wellbeing in four East and Southeast Asian societies who share Confucian tradition (Inoguchi and Shin 2009), based on an original cross-national survey which was conducted by the authors' team.

The survey "International Comparative Surveys on Lifestyle and Values" was conducted in Japan, Korea, Taiwan by web and in Vietnam by face-to-face interview from 2015 till 2017 based on a standardized questionnaire. The populations were individuals aged 20–69 in Japan, Korea, and Taiwan, and aged 18 and over in Vietnam. Nationwide quota samples being proportionated to a census distribution were drawn from the monitors of web-survey agents in Japan, Korea, and Taiwan, and by area sampling in Vietnam. The number of cases ranges from 1,200 to 11,786. Refer to Table 1 for further details of the survey.

Though existing literature has pointed out a micro-level correlation between religion or spirituality in general and subjective wellbeing as we mentioned earlier, there have been no research and surveys focusing exactly on "ancestor worship" so far. Thus, we tried to establish instruments for measuring ancestor worship.

Manabe (1999) provides a sociological definition of ancestor worship. She argues that ancestor worship is mediated by religious or ceremonial systems for honoring deceased ancestors and invoking their protection. It involves a reciprocal relationship between descendants who pay tribute to ancestral spirits and the ancestral spirits themselves who bestow their protection upon the descendants. She further argues that this relationship encompasses the aggregate of the systems of doctrinal belief, ceremonial practice, and customary behavior. Ikeoka (1999) defines religious services for ancestors as a set of institutionalized formalities directed toward tombs, gravestones, memorial tablets and portraits of the deceased, prayers, offerings, or offerings of food to the deceased. We will limit the scope of our inquiry to formalities for mourning ancestors, funeral rites that involve the living switching places with the deceased, and the use of graves to which the deceased are interred.

Table 1 Survey design: international comparative surveys on lifestyle and values

	Japan	Korea	Taiwan	Vietnam
Timing	Feb 18–24, 2015	Jul 14–22, 2015	July 1–31, 2017	July–December 2015
Mode	Web	Web	Web	Face-to-face
Sample	20–69 Quota by gender, age, city size, and region from monitors	20–69 Quota by gender, age, and region from monitors	20–69 Quota by gender, age, city size, and region from monitors	18 and over Area sampling in 6 administrative units
Eff. cases	11,786	2,000	2,303	1,200

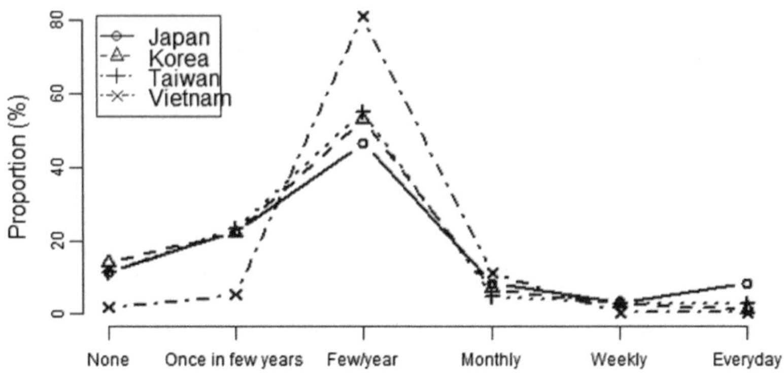

Fig. 1 Distribution of "frequency of attending memorials of the deceased". *Source* International Comparative Surveys on Lifestyle and Values

Considering these definitions, we finally operationalized ancestor worship by the following two instruments, i.e., behavior and attitude. As a behavior linked to ancestor worship, we asked frequency of conducting ancestral rituals or memorials in everyday life. The question was "How frequently do you attend memorials or services for those close to you or ancestors who have already died?" The options were "Not at all", "Once every few years", "Few times a year", "Monthly", "Weekly", and "Almost every day". Figure 1 indicates the distribution of frequencies in the four societies. Japan, Korea, and Taiwan share similar distributions whereas Vietnam has a distribution concentrating "Few times a year." The reason is that Vietnamese people normally perform ancestor worship within four generations, and the worshiping days are set for each passing person once a year, on the day they passed away.

As a measure of attitude toward ancestor worship, we asked respondents to what extent they think their descendants will worship them after they pass away. This question was taken because the function of ancestor worship as the basis of the self stems from a reciprocal relationship between ascendants and descendants in an imagined family lineage, namely the inter-generationally transmitted rolls by which the living worships the deceased (= ascendants) and the deceased in turn protects the living (=descendants) spiritually. The instrument was "Do you think your children or descendants will hold memorials or services for you in the future after your passing?", and the options were "Definitely not", "Probably not", "Maybe", "Probably", and "Definitely". Figure 2 shows the distribution of responses. Here again Japan, Korea, and Taiwan share similar distributions whereas Vietnam has a unique distribution that more than half (=54%) believe they will "definitely" be worshiped by their descendants.

Subjective wellbeing was measured by subjective happiness on an eleven-point scale from 0 (Very unhappy) to 10 (Very happy). The instrument was "How happy are you currently?" Figure 3 shows the distributions and mean scores of the four societies. Again, three East Asian societies have same patterns and Vietnam is different from

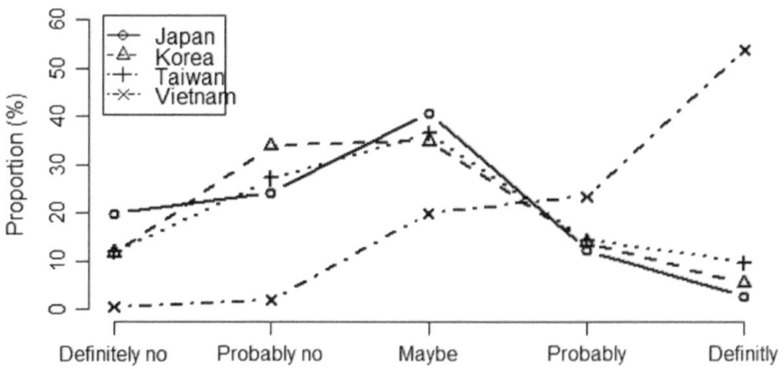

Fig. 2 Distribution of "expectation to be worshiped by descendants". *Source* International Comparative Surveys on Lifestyle and Values

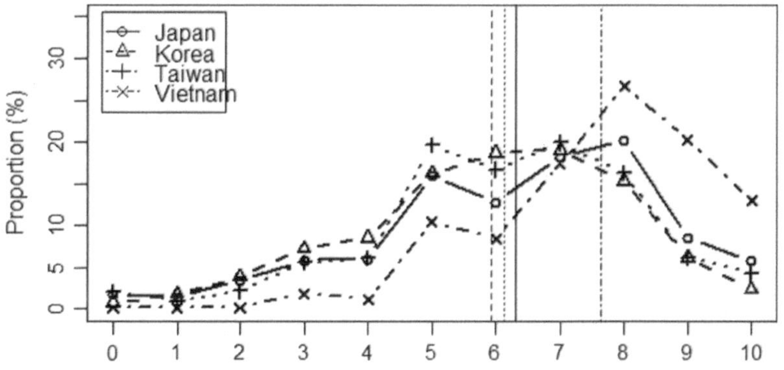

Fig. 3 Distributions and means of subjective happiness. Vertical lines show the mean scores in each society. *Source* International Comparative Surveys on Lifestyle and Values

them. The mean scores of subjective happiness are relatively lower in East Asian societies (Japan 6.3; Korea 5.9; Taiwan 6.1) than in Vietnam (7.6).

In summary, the average level of both of ancestor worship and subjective wellbeing is lower in Japan, Korea, and Taiwan, and higher in Vietnam despite that all the four societies share the Confucian tradition.

Then, how are the correlation between ancestor worship and subjective wellbeing in these four societies? Figure 4 shows the association between practice of ancestor worship and subjective wellbeing in these four societies by plotting mean scores of subjective happiness for each category of the "frequency of attending memorials" question. Though again Japan, Korea, Taiwan share similar patterns and Vietnam has

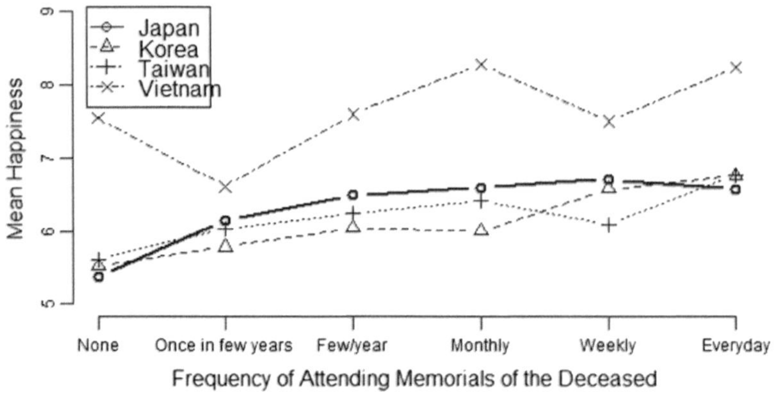

Fig. 4 Association b/w practice of ancestor worship and subjective wellbeing. *Source* International Comparative Surveys on Lifestyle and Values

a unique pattern, those who attend memorials more frequently have higher level of subjective wellbeing in all the four societies.[1]

Figure 5 presents how the other component of ancestor worship, namely expectation to be worshiped by descendants associates with subjective wellbeing. Here we see accustomed patterns between the four societies, but in all the societies, expectation to be worshiped has a positive association with subjective wellbeing.[2]

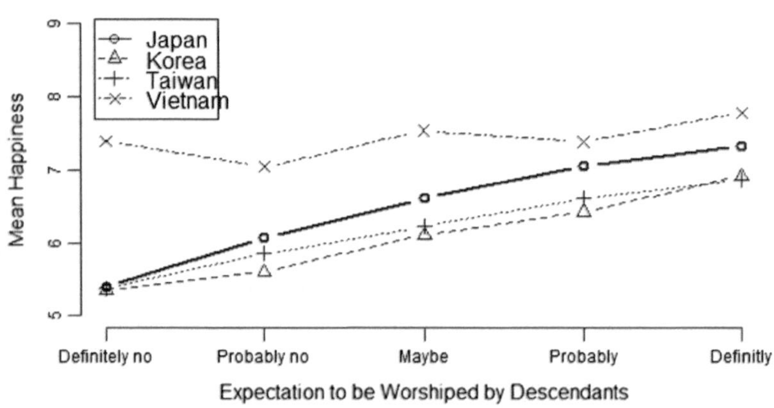

Fig. 5 Association b/w expectation of ancestor worship and subjective wellbeing. *Source* International Comparative Surveys on Lifestyle and Values

[1]OLS regressions of subjective happiness to worship practice controlling for gender, age, marital status, years of education, employment, and log equivalent income also revealed significant positive effects of worship practice (as a categorical variable) on wellbeing in all the four societies.

[2]OLS regressions of subjective happiness to worship expectation controlling for the variable above also revealed significant positive effects of worship expectation (as a categorical variable) on wellbeing in all the four societies.

So far, we have confirmed that behavior and attitude toward ancestor worship has a positive micro-level effect on subjective wellbeing in the four societies who share Confucian tradition. This finding suggests the belief in the traditional custom of ancestor worship would still promote one's wellbeing even in the advanced economies such as Japan, Korea, and Taiwan.

However, all these Asian societies including Vietnam have experienced rapid modernization and changes in population and family structure in these decades. Would such social change undermine people's traditional values and life style to derive new types of the self and interconnectedness in these societies? The following sections address this question by historical and qualitative methods.

2 Ancestor Worship and Its Transformation in Japan and Vietnam

This section overviews the transformation of ancestor worship in Japan comparing with Vietnam. The reason we take Vietnam as a reference is in our quantitative findings in the previous section. Since Vietnam is in a unique position among the four Confucian societies regarding the level of ancestor worship and wellbeing as well as the association between them, comparing Japan with Vietnam would give us some hints about the effect of modernization on the changing practice and attitude.

2.1 Transformation of Ancestor Worship and Funerals in Japan

Japan has experienced a drastic change in population and family structure since the Second World War. Along with a steady decline in the proportion of junior population under the age of 14,[3] the proportion of senior citizens aged 65 and above has increased gradually to reach 26.3% in 2015, which is the highest score in the world. Figure 6 shows the changes in age composition from 1950 to 2015 in Japan.

The composition of household types has also changed since the 1980s (Fig. 7). Though the proportion of nuclear family (two-generation households with parents and children) remains stable at around 55–60% during the past four decades, the proportion of three-generation households (the stem-family system that characterizes traditional Japanese households, *ie*) has declined rapidly from 15% in 1980 to only 4% in 2015. In accordance with this trend, one-person households have increased from 20% in 1980 to 35% in 2015. About one third of these one-person households belonged to a person aged 65 and above in 2015.[4]

[3]Total fertility rate has decreased from 2.96 in 1950 to 1.48 in 2015.

[4]5,927,686 out of 17,375,543 equals 34.1% (2015 Population Census).

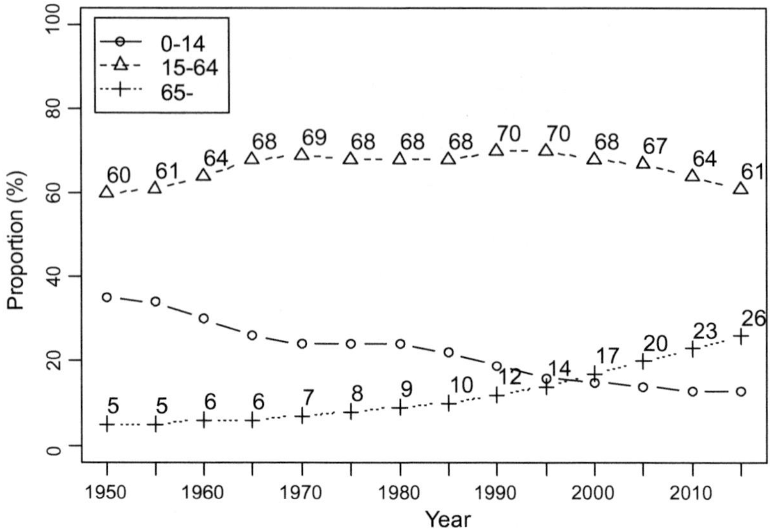

Fig. 6 Trend in age composition in Japan. *Source* UN World Population Prospects: The 2017 Revision

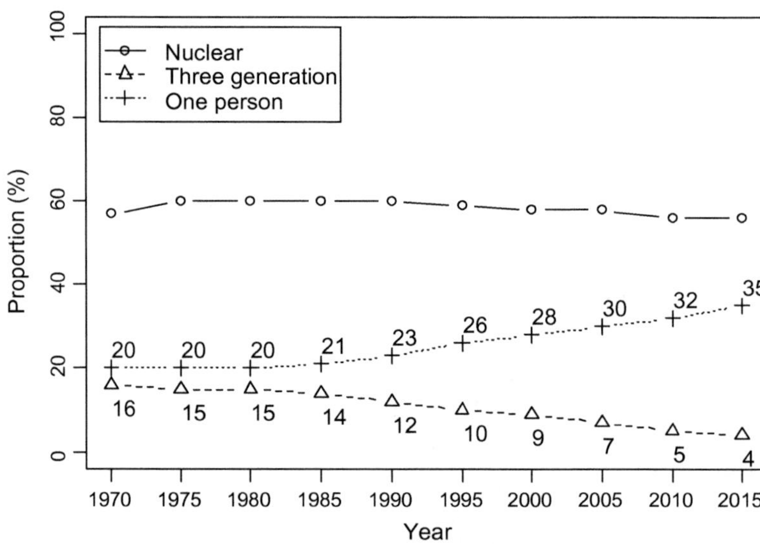

Fig. 7 Trend in family types of households in Japan. *Source* Population Census

Along with these changes in population and family structure as well as population shifts from rural to urban areas, people's attitudes toward ancestor worship have also transformed. The Japanese National Character Survey, a repeated national representative survey conducted by the Institute of Mathematical Statistics every five years since 1953, has asked respondents whether they would adopt a son if they have no sons. Allowing non-kin adoption in order to perpetuate households without sons is a typical feature of the Japanese traditional household system, which is different from other Confucian societies where preserving patriarchic lineage is the most important value. Figure 8 shows that those who would adopt a child was about three fourth (74%) in 1953 but fell into about half (36%) for only two decades, finally hitting 20% in 2013. In contrast, the proportion of those who would not adopt a son has increased, from only 16% in 1953 to over half (57%) in 2013.

These drastic changes in population structure and attitude towards the traditional household system would have caused the decline in ancestor worship during these decades. Exactly to say, the scope of ancestor worship has narrowed significantly, so that it now involves nothing more than mourning for one's departed kin. The decline in birthrates brought about by higher income (Komine and Japan Center for Economic Research 2007) and changes in family structure (Ochiai 1997; Ueno 2016) have rendered bloodline-centered (*ie*-type) ancestor worship unfeasible.

Mori (2014) identified two stages in the process of change in ancestor worship and funerals. First, the rise of civil society decreased individuals' dependence on the local community. As a result, local communities have disengaged from their supportive role in funerals, and funerals have become increasingly commercialized. He calls

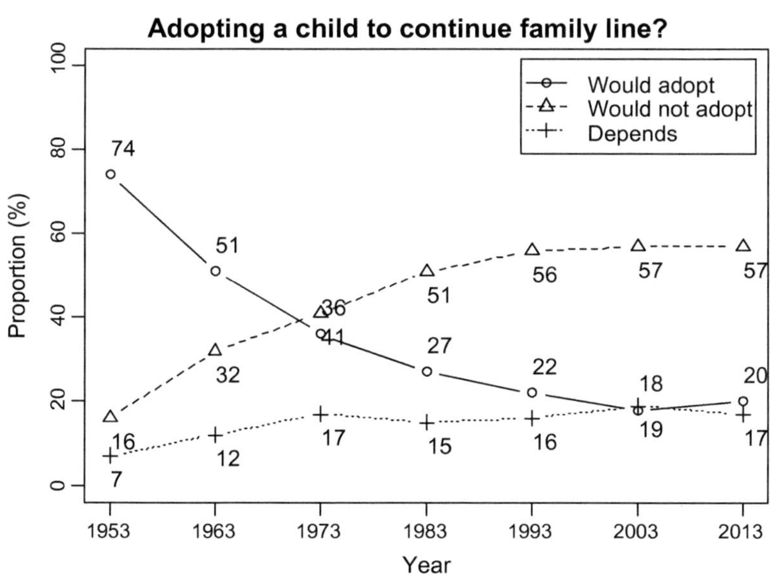

Fig. 8 Changes in the attitude on adopting a child. *Source* The Japanese National Character Survey, The Institute of Mathematical Statistics

this stage as "first modernization." Even during the period of high economic growth, individualization of funerals was limited to some extent by the Japanese-style modern family structure, which had subsumed the role of ancestor worship. However, this funerary system eventually started breaking apart in the 1990s, marking the beginning of Japan's "second modernization," under which funerals became increasingly commercialized and diverse. Shimane (2012) argued that even co-workers, who had substituted the role of the local community, started losing the auxiliary role they had played in funerals owing to the shrinking and aging population.

Inoue (2003) discussed the transformation in family structure and trends in ancestor worship in terms of changes in burial practices. She described the traditional stem family as follows:

> The archetypal Japanese stem family was characterized by the practice of paying homage to one's ancestors --- the forbears in one's family line --- and the practice of intergenerational handovers (handing over the family headmanship and ritual authority to the next generation) in the expectation of, and in order to facilitate, the perpetual continuity of such handovers across successive generations (genealogical continuity) (Inoue 2003: 3).

According to Inoue, the key characteristic of the modern-day nuclear family is that it has lost its congruity with this stem family system. The natural consequence of this development is that ancestor worship and funeral rites have come to be conducted exclusively within the conjugal family unit, and the broader kinship community is increasingly disengaging from them.

These changes have led to the growing popularity of "family funerals" in Japan. Only close kin and friends of the deceased attend family funerals. The notion of "family" in such funerals is very specific and does not match the notion of the extended family or nuclear family; it refers exclusively to the "family" that is subjectively envisaged by the people conducting the ancestor worship or funeral. Therefore, even though members of one's family of orientation are technically family, there is an emerging trend of excluding one's siblings from ancestor worship and funeral rites.

The family funerals that one of the authors attended on 2007 and 2017 had only around 20 to 30 attendees. Taking into consideration the fact that the families concerned were socially upstanding and well off financially, and the fact that the funerals for previous generations were much grander affairs, these funerals seem to illustrate the scale of the changes that have taken place since the end of the 20th century. If attendance at funerals is so limited, then one would expect future occasions of ancestor worship to receive even fewer attendants, barring some unforeseen cultural shift.

The phenomenon occurring in the 2010s is even more alarming. The shrinking and aging population and rising numbers of unmarried people are beginning to further undermine ancestor worship and funeral rites that nuclear families have hitherto conducted. For instance, Yokosuka city in Kanagawa prefecture recently started an end-of-life/after-death support program. They explained the reason for doing so as follows:

> Currently, the number of people in Yokosuka who are living alone stands at 10,000 and is increasing. Every year, there are around 50 deaths for which no one comes forward to

> handle the burial arrangements despite the identity of the deceased being known. Welfare commissioners and members of neighborhood associations have expressed the opinion that we should intervene early in concerns related to the end-of-life care of elderly people who live alone (Yokosuka City website).

A report broadcasted by NHK (Japan Broadcasting Corporation) in 2010 motivated these actions. In the report, NHK coined the term "dying in isolation" (*koritsu-shi*) to describe people who end their lives alone with no one to look after them and announced that an estimated 32,000 people a year die in isolation (NHK 2010).

In this way, modernization in Japan has led to community's disengagement from ancestor worship and funerals, and this in turn has severed connections among neighbors, co-workers, friends, acquaintances, and even in some cases, among family members. Community's disengagement from ancestor worship is thus a crystallization of a larger process by which people are deprived of their bonds with the deceased and forced them to live unstable and atomized lives bereft of a sense of connection with the past or an outlook for the future.

Kotani (2017) claims that modernization-driven transformation of ancestor worship and funerals are not exclusive to Japan:

> The Taiwanese have traditionally adhered to a spirit of mutual aid, in which kin and patrilineal relations help each other. However, shrinking and aging population, longer life expectancy, and nuclearization of family are progressing apace. Consequently, new social problems are emerging, including the limitations of care within the family and cases of elderly people dying alone (Kotani 2017: 146).

In other words, after a society undergoes a period of growth and the demographic dividend drives up living standards, a change in family structure inevitably follows, which will in turn transform funerals and ancestor worship.

In the next sub-section, we compare Japan with Vietnam, where despite drastic socioeconomic changes, people try to retain the traditional ancestor worship and funerary rites through strong patrilineal bonds. We will focus on the examples of ancestor worship and funeral rites among the Kinh people, as their social structure resembles that of Japanese farming communities.

2.2 Transformation of Ancestor Worship and Funerals in Vietnam

We can see from Fig. 9 that the population structure in Vietnam features high proportion of working ages (70% in 2015) and relatively low proportion of aged population (only 7% in 2015), despite decrease in fertility rate (from 5.40 in 1950 to 1.96 in 2015). Though some of these features are caused by the high mortality rate during the Vietnam War (1955–1975), current population structure in Vietnam is still younger than it is in Japan. However, the average household size in Vietnam has gradually decreased, as it has in Japan (Fig. 10). Therefore, we would expect Vietnam to follow the Japanese trajectory in the near future.

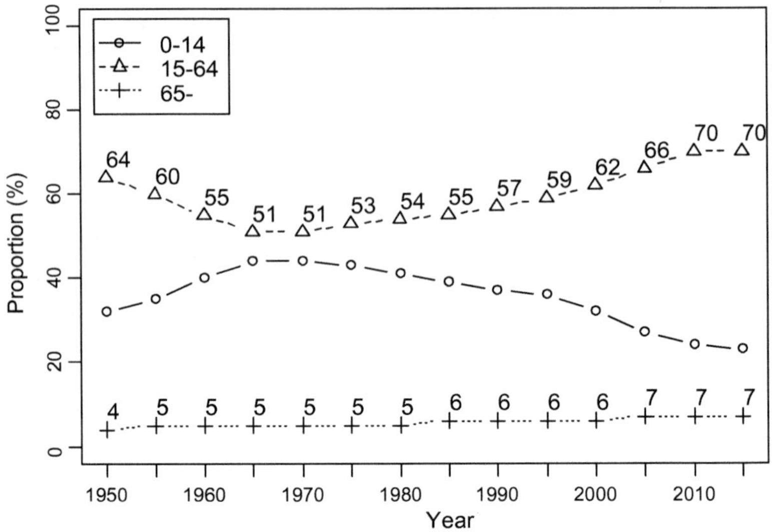

Fig. 9 Trend in age composition in Vietnam. *Source* UN World Population Prospects: The 2017 Revision

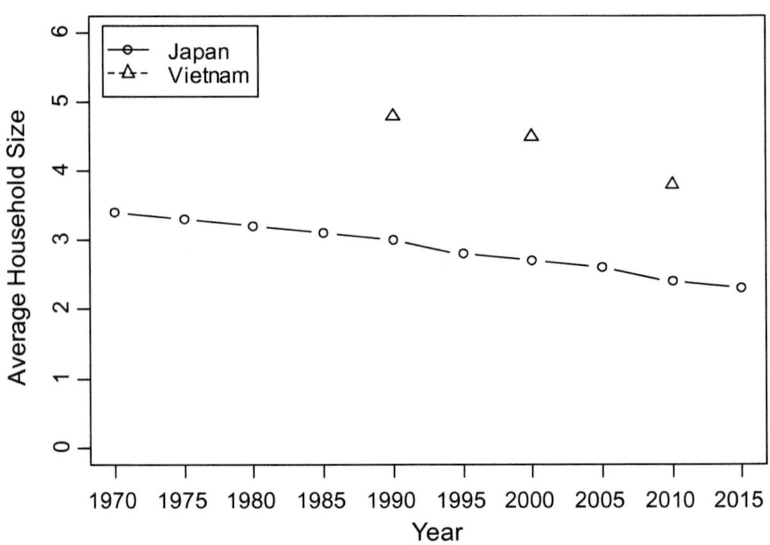

Fig. 10 Trend in average household size in Japan and Vietnam. *Source* Population Census in Japan and Vietnam

Among the Kinh people, which comprise the majority of the population in Vietnam, paternal lineage is prioritized. Children take their father's family name, patrilocal residence after marriage is popular, and inherited assets are normally allocated to sons. The oldest son or any son who finally co-resides with his parents has primary responsibility for caring for the parents until their death and worshiping their ancestors (Phuong and Dang 2016). Sons are the heirs to the lineage, helping to extend bonds between the living and the dead, and to bridge living generations with their ancestors.

Funerals are major life-cycle events and are typically open to attendants other than immediate family members. A decent funeral is one that is organized among family members, relatives, friends, co-workers, neighbors, and acquaintances. People feel the responsibility to come to a funeral to offer incense and pay the final tribute to the deceased because, as a proverb says, "everything is forgiven when it comes to rest." The success of a funeral speaks much of the deceased and his or her families in the eyes of their community. The more people that attend a funeral, the more prestige the deceased and the family have (Dang 2015: 230). By attending a funeral ceremony, people can set up new social networks or consolidate their existing social cohesion. The funeral family, therefore, usually holds a meal to express their gratitude to the guests as well as to show that they have a good social life. Funeral ceremony is thus a place where the bond between the deceased and the living is established.

Vietnamese people believe the soul of a dead person continues to be present among their descendants on earth for many generations. Maintaining bonds with the deceased is increasingly popular and becoming a unifying aspect of Vietnamese culture. At its extreme, ancestor worship may even be considered a religion (Van 2001), as regardless of their religious affiliation, almost every Vietnamese has an ancestor altar in their home and practices worship quite frequently. The intensification of ritual practicing of ancestor worship was recognized since the Doi-Moi period (Luong 1993) in the restoration of ancestor worship, genealogy reproduction, temple of forefathers building, and so on. The offerings and practices of ancestor worship are done frequently during important household events, such as the starting of a new business, a newborn child, a new house, and so on.

Ancestor worship moreover functions as a kinship-building activity for living family members. The anniversary of a relative's death is an opportunity for family members to gather together to recall their ancestors, demonstrate their sincerity, and organize feasts, all of which strengthens solidarity among them. In this way, the anniversary of a deceased builds and expands social network and capital among the living.

Nowadays in Vietnam we witness the dual trends of ancestor worship, with on one side the restoration of traditional practices and honoring kinship activities, and on the other side the personalization of ancestor worship where such practices were overly specialized as a family religion. There are still households respecting the custom of organizing ancestor worship among extended family members; however, in many other families, the anniversary of the deceased is no longer celebrated with a reunion. Instead, each member of the extended family commemorates the death of their ancestors in their nuclear family. The pattern of community worship has thus turned into one of individual worship.

2.3 Similarities and Differences in Ancestor Worship Between Japan and Vietnam

Japanese and Vietnamese societies (particularly the Kinh people in the north of the country) have shared many similarities in terms of family system, ancestor worship, and funerary customs. One common fact that explains why the two societies are closely linked might be wet-rice farming. As Dawkins (1976) argued, not to split farmland but to concentrate it to a head family is "the most adapted behavior," because dividing it into small portions would lower productivity of farming. Branch families must have subordinate to a head family, which in turn protected branch families. Ancestor worship has worked as a symbol to unify enlarged families under a head family (Takeda 1976).

It is noteworthy that Japanese ancestor worship and funerary customs, traditionally regarded as firmly rooted in society, faded away (at least in urban areas) over a mere 40 years as a result of high economic growth beginning in the mid-1950s and changes in family structure and lifestyle. Vietnam has also experienced significant social changes commensurate with its economic growth (Hashimoto 2017; Itagaki 2008; Sato 2017). Given its shrinking population, it is uncertain whether Vietnam will continue to retain its ancestor worship and the family structure that underlies it.

3 Bonds with the Living Emerging from Bonds with the Deceased

Japanese civil law, which was enacted in 1899, established the traditional household system, *ie*, as a formal institution. In *ie*, authority over the household estate was concentrated in the household head. After the Second World War, the *ie* was legally repudiated; however, it lived on in people's attitudes, and ancestor worship and funerals continued for many years to be entrusted to patrilineal descendants primarily the eldest son. The shrinking and aging of the Japanese population has made it unfeasible to maintain this practice, as it is becoming difficult to transfer such practice to the next generations, let alone to a male heir. This situation resulted in rapid changes from the 1990s onward in funerals and the system of family grave inheritance. These changes in turn have transformed norms in everyday rituals and how people expect to be treated after death.

In the *ie*-centered society, attitudes and behaviors were strictly restricted by traditional norms. The upside of this was that people were able to derive a sense of purpose and meaning of life from a "perpetual" continuity between ancestors and descendants. The process of modernization and industrialization has made people's lives more fluid, and people have severed traditional blood and neighborhood ties and embraced a new social network: the modern-day workplace. Japanese companies provide a source of community in itself: the community of co-workers. However, as population ageing extends a post-retirement period of individuals, the workplace has become just one facet of the life course.

This applies to men as well as women. The rise of the "new family" after the Second World War liberated women from the legal *ie* system, under which many women located the meaning and purpose of their lives in conjugal nuclear families. The modern nuclear family does not have the same perpetual continuity as it did under the *ie*. In many cases, children of nuclear families prefer living apart to living in two-generation households, against the expectation or will of their parents (Tsai and Yang 2017). Parents themselves, assured by well-established pension and social security systems, feel increasingly comfortable with getting old together after their children have left the home.

Along with this transformation, many people are becoming worried about lacking burial places and about having someone to look after their burial places. As long as funerals are conducted based on the stem family structure, those who have no successors, such as singles or the divorced, are refused a place in graveyards. Since Buddhist temples or graveyard managers give the deceased the right to occupy the burial space "perpetually" with the expectation that their successors will make donations and pay the costs of maintenance, they are reluctant to give such right to those with no prospect for remuneration. Therefore, single persons, especially unmarried or divorced women who do not have the right to inherit the family grave, have little chance of securing a burial space or getting someone to look after it.

Faced with this, some individuals have tried to create communal graves or communal burial spaces in search of new human relationships. Haruyo Inoue, who we cited earlier, worked with *Myōkō-ji* Temple in Niigata Prefecture to find a solution to this difficulty. In 1989, they opened the "tomb of repose" (*annon-byō*) within the temple ground. They intended to overturn the traditional grave system, and to explore a new type of communal grave that are available to single women. In line with this idea, Inoue also advocated to expand the practice of "cherry blossom burials" (*sakura-sō*) nationwide. In cherry blossom burials, a single grave is created under a cherry tree. The grave is composed of multiple adjacent spaces that each deceased has the right to use. This configuration of the grave induces a new kind of relationship between those who are to be buried in the grave after their death. This new relationship, "burial-friendship" (*haka-tomo*) (Inoue 2012), is a mutual bond that transcends family boundaries. Those who are strangers to each other but will be buried in a neighboring plot in the future, gather regularly and get to know each other as "grave-friends."

Inoue (2012: 81) presents a life history of a woman in her sixties as a case:

> From my infancy, I never wanted to be bound in the Japanese traditional household system (*ie*) and to be sealed within a stone-hardened grave. I was born in Tokyo and lost my father to illness at the age of 15. A Buddhist monk of the *Bodai-ji* Temple in his hometown travelled

> to Tokyo to conduct the funeral service. My father was buried in the family tomb in his hometown. However, due to my family's complicated situation, I felt no comfort in the pomp funeral ceremony, the multitude of attendees, the vast gravesite, and the large tombstone.

After the funeral, the woman became a Christian, met her future (now late) husband in church, and started a family of her own. Her late husband was interred in the church's communal graveyard, and the woman initially had no doubt about being buried there herself after she dies. However, the woman subsequently left the church and thus lost her emotional place-of-belonging. It was at this point that she found out about the cherry blossom burials. When the woman learned there was such a burial space in her neighborhood, she attended a briefing meeting. The woman started attending regular *haiku* meetings with her grave-friends. The woman sums up this experience as follows: "Cherry blossom burials gave me not only a secure place for my afterlife but also a power to live which enriched "the present" of my life while alive. I sincerely appreciate this relation." (Inoue 2012). This new trend in the form of burial suggests that the declining feasibility of perpetuating the traditional household system and ancestor worship has induced a new form of "bonds with the living" in contemporary Japanese society.

In traditional ancestor worship, those who are on the patrilineal family tree but lived in the past that the living does not know them still supported a person's identity. In other words, the "bonds with the deceased" were assured by the traditional imagination to communicate with "the abstract dead". Modernization has deprived such abstract imagination. Some foreign anthropologists who investigated Japanese ancestor worship and funerals have already pointed out the tendency of shrinking imagination in the 1970s (Plath 1980; Smith 1974). The shrinking imagination has finally resulted in a collective of others of whom one knows whether or not they are still alive (Suzuki 1998, Suzuki 2013). The "bonds with the living" in contemporary Japan are a logical consequence of this trend, in which living others substitute the role of ancestor worship as a basis of the self.

4 Conclusion

We can now answer the research questions in this chapter, that is, how Japanese people cope with the current situation in which the traditional assurance for their afterlife does not work well due to changing population and family structure, and to identify the ways in which they seek alternative bases for their identity.

Our answer to these two questions is that the bonds with the deceased of whom people directly knew and with living individuals outside the immediate family substitute vanishing bonds with the more abstract notion of ancestors. Though this new trend is often observed in those marginalized by the traditional patrilineal household system, such as unmarried or divorced women, it is expected to spread to other people, given persistent low fertility in Japan. This silent transition from "bonds with the deceased" to "bonds with the living" as a basis of the self can be regarded as a Japanese style of adaptation to the structure of post-material modernity.

As we mentioned in the introduction of this chapter, anthropologists and cultural psychologists have discussed the contrast in the concept of the self between Western and East Asian culture as "independence" in the West versus "interdependence" in the East. In Euro-American, Judeo-Christian culture, the basis of the self is in principle assured by a direct connection between the God and the individual. This strict idea of the relation between the God and the individual, especially established by Calvinism, severely threatened pious people in early modern society who had a strong interest in the relief of their souls at Doomsday. Though the solutions to this anxiety have shifted through the secularization of the society, modern people living under the Western concept of the self inevitably continue to suffer from this anxiety (Maki 1981).

In contrast, East Asian culture accentuates interdependence of the self, the connectedness of the self to others. Among many scholars across the globe discussing this topic, Eshun Hamaguchi named this feature of Japanese society "contextualism" (*kanjin-shugi*). He identified the three features of contextualism as interdependence, mutual trust, and regarding human relations as an end instead of as a means (Hamaguchi 1977). The reciprocal nature of traditional ancestor worship based on the mutual trust between ascendants and descendants certainly fits this feature of Japanese society. Logically speaking, however, ancestor worship is not the only way to assure people's self-identity. Interdependence and mutual connection with the living is another way for people to imbue meaning into their lives.

Though the transition from "bonds to the deceased" to "bonds to the living" is not yet complete, we can expect this new trend will develop in the near future both in Japan and Vietnam due to the irreversible changes in the population structure. Following the trajectory of this trend will be an important task for future studies.

Acknowledgements This work was supported by the MEXT-Supported Program for the Strategic Research Foundation at Private Universities of Japan, 2014–2018 (S1491003) and by a research grant from the Senshu University Research Fellowship 2017. The "International Comparative Surveys on Lifestyle and Values" were designed and conducted by the Center for Social Well-being Studies, Institute for the Development of Social Intelligence, Senshu University, Japan, in collaboration with Social Well-being Research Consortium in Asia.

References

Dawkins, R. (1976). *The selfish gene*. New York: Oxford University Press.

Hamaguchi, E. (1977). *Nihon rashisa no saihakken* [Re-discovering Japaneseness]. Tokyo: Nihon keizai shinbunsha.

Hashimoto, K. (2017). *Ushinawareru shikuro no shita de* [Under the disappearing cyclos: Vietnamese society and history]. Tokyo: Harvest.

Ikeoka, Y. (1999). Sosen saishi [Ancestral rites]. In Y. Shoji (Ed.), *Fukushi shakai jiten* [Encyclopedia of welfare society] (pp. 650–651). Tokyo: Kobundo.

Inglehart, R. F. (2010). Faith and freedom: Traditional and modern ways to happiness. In E. Diener, J. F. Helliwell, & D. Kahneman (Eds.), *International differences in well-Being* (pp. 351–388). Oxford: Oxford University Press.

Inoguchi, T., & Shin, D. C. (2009). The quality of life in Confucian Asia: From physical welfare to subjective well-being. *Social Indicators Research, 92*(2), 183–190.

Inoue, H. (2003). *Haka to kazoku no henyo* [Transformation of graves and families]. Tokyo: Iwanami-shoten.

Inoue, H. (2012). *Sakura-so* [Cherry blossom graves: Dreams for sleeping under cherry trees]. Tokyo: Sanseido.

Itagaki, A. (2008). *Betonamu* [Vietnam: Transforming medical treatment and rituals]. Yokohama: Shunpu-sha.

Komine, T., & Japan Center for Economic Research. (2007). *Cho-choki yosoku oiru ajia* [Hyper-long-term prediction for aging Asia: Changing world population and economic maps]. Tokyo: Nihonkeizaishinbunsha.

Komoto, M. (2001). *Gendai nihon ni okeru senzo saishi* [Ancestor worship in contemporary Japan]. Tokyo: Ochanomizu Shobo.

Kotani, M. (2017). *Hitori-shi jidai no ososhiki to ohaka* [Funerals and graves in the age of dying alone]. Tokyo: Iwanami-shoten.

Kumon, S. (1982). Some principles governing the thought and behavior of Japanists (contextualists). *Journal of Japanese Studies, Seattle, 8*(1), 5–28.

Luong, H. V. (1993). Economic reform and the intensification of rituals in two north Vietnamese villages, 1980–90. In B. Ljunggren (Ed.), *The challenge of reform in Indochina* (pp. 259–291). Cambridge, MA: Harvard Institute for International Development, Harvard University.

Maki, Y. (1981). *Jikan no hikaku shakaigaku* [Comparative sociology of time]. Tokyo: Iwanami shoten.

Manabe, Y. (1999). Sosen suhai [Ancestor worship]. In Y. Shoji (Ed.), *Fukushi shakai jiten* [Encyclopedia of welfare society] (pp. 662–663). Tokyo: Kobundo.

Markus, H. R., & Kitayama, S. (1991). Culture and the self: Implications for cognition, emotion, and motivation. *Psychological Review, 98*(2), 224.

Marques, S. C., Lopez, S. J., & Mitchell, J. (2013). The role of hope, spirituality and religious practice in adolescents' life satisfaction: Longitudinal findings. *Journal of Happiness Studies, 14*(1), 251–261. https://doi.org/10.1007/s10902-012-9329-3.

Mori, K. (2014). *Haka to soso no yukue* [The future of graves and funerals]. Tokyo: Yoshikawa-kobundo.

NHK. (2010). *Muen shakai* [Society without connectedness]. Tokyo: Bungei-shunju.

Ochiai, E. (1997). *The Japanese family system in transition: A sociological analysis of family change in postwar Japan*. Tokyo: LTCB International Library Foundation.

Dang, T. V. P. (2015). *The collective life: The sociology of voluntary associations in North Vietnamese rural areas*. Hanoi: National University of Hanoi Press.

Phuong, P. T. T., & Dang, T. V. P. (2016). Vietnam, families in. In C. L. Shehan (Ed.), *The Wiley Blackwell encyclopedia of family studies* (Vol. 4 Q-Z, pp. 2022–2026). Chichester: Wiley.

Plath, D. W. (1980). *Long engagements: Maturity in modern Japan*. Stanford, CA: Stanford University Press.

Sato, K. (2017). *Hen-yo suru betonamu no shakai kozo* [Transformation of Vietnamese social structure]. Tokyo: Senshu University Press.

Sawatzky, R., Ratner, P. A., & Chiu, L. (2005). A meta-analysis of the relationship between spirituality and quality of life. *Social Indicators Research, 72*(2), 153–188. https://doi.org/10.1007/s11205-004-5577-x.

Shimane, K. (2012). The experience of death in Japan's urban societies. In N. Aveline-Dubach (Ed.), *Invisible population: The place of the dead in East Asian megacities* (pp. 29–49). Lanham: Lexington Books.

Smith, R. J. (1974). *Ancestor worship in contemporary Japan*. Stanford, CA: Stanford University Press.

Suzuki, H. (1998). Japanese death rituals in transit: From household ancestors to beloved antecedents. *Journal of Contemporary Religion, 13*(2), 171–188. https://doi.org/10.1080/13537909808580829.
Suzuki, I. (2013). Beyond ancestor worship: Continued relationship with significant others. In H. Suzuki (Ed.), *Death and dying in contemporary Japan* (pp. 141–156). New York: Routledge.
Takeda, C. (1976). *Nihonjin no ie to shukyo* [Japanese household and religion]. Tokyo: Hyoronsha.
Tsai, M. C., & Yang, W. S. (2017). Familial exchange and intergenerational contact in East Asian societies. In M. C. Tsai & W. S. Yang (Eds.), *Family, work and wellbeing in Asia* (pp. 21–42). Dordrecht: Springer.
Tu, W.-M. (1996). *Confucian traditions in East Asian modernity: Moral education and economic culture in Japan and the four mini-dragons*. Cambridge, MA: Harvard University Press.
Uchida, Y., Norasakkunkit, V., & Kitayama, S. (2004). Cultural constructions of happiness: Theory and empirical evidence. *Journal of Happiness Studies, 5*(3), 223–239. https://doi.org/10.1007/s10902-004-8785-9.
Ueno, C. (2016). *Onna tachi no sabaibaru sakusen* [Survival strategy of women]. Tokyo: Bungeishunju.
Van, D. N. (2001). *Dân tộc Văn hóa Tôn giáo* [Ethnics, culture, and religion]. Hanoi: Social Sciences Publisher.
Yum, J. O. (1988). The impact of Confucianism on interpersonal relationships and communication patterns in East Asia. *Communications Monographs, 55*(4), 374–388.

Masayuki Kanai is Professor of Sociology at School of Human Sciences, Senshu University, Japan. He is Co-Vice President of the International Society for Quality of Life Studies, Secretary-Treasurer of Research Committee on Rational Choice (RC45), and a board member of Research Committee on Social Indicators (RC55) of the International Sociological Association. His current research project focuses on social mechanisms of subjective wellbeing in comparative perspective. His articles have appeared in *Sociological Theory and Methods*, *Senshu Social Well-being Review*, among others. His recent book includes *Contemporary Japanese Sociology* (coeditor, Sage, 2018) (mkanai@senshu-u.jp).

Katsumi Shimane is Professor of Sociology, Dean of School of Human Sciences, Senshu University, Japan. He is focusing his research to modernization of funeral ceremony compering diverse societies. His articles have appeared in *Philosophical Transactions of the Royal Society B*, *The Monthly Bulletin of Social Science*, *Annuaire de la Société japono-française de Sociologie*, *Senshu Social Capital Review*, among others. His book chapters include, "The Experience of Death in Japan's Urban Societies", *Invisible Population; the place of The Dead in East Asian Megacities* (Lexington Books, 2012) (shimane@isc.senshu-u.ac.jp).

Dang Thi Viet Phuong is researcher at the Institute of Sociology, Vietnam Academy of Social Sciences, and inviting lecturer of Vietnam-Japan University. Her current research projects focus on Vietnamese rural families in change and civil society. Her articles appeared in *Journal of Current Southeast Asia Affairs*, *Vietnam Social Sciences Review*, *Senshu Social Well-being Review*, *Senshu Social Capital Review*, among others. Her recent book and book chapters include "Vietnam", in Akihiro Ogawa (ed.), *Routledge Handbook of Civil Society in Asia* (Routledge, 2017, co-author); "Vietnam, families in", in Constance L. Shehan (ed.), *The Wiley Blackwell Encyclopedia of Family Studies* (Blackwell, 2016, co-author). "*The collective life: The sociology of voluntary associations in North Vietnamese rural areas*" (National University of Hanoi Press, 2015) (dangvietphuong@ios.org.vn).

Intimacy, Similarity, and Equality Among Married People in East Asia

Ming-Chang Tsai

Abstract This paper endeavors to examine personal intimacy by way of comparative investigation of the confiding behaviors of married men and women in East Asian societies. A number of hypotheses were derived from assortative marriage and pure relationships theories for empirical testing. Assortative marriage contends that similarity is a precondition to intimate disclosure, while pure relationship theory proposes equality to be a major determinant. Using data from the East Asian Social Survey collected in Taiwan, South Korea, Japan and China, this study finds that educational and age assortative mating do not facilitate confiding of husbands and wives. In contrast, power sharing in decision-making produces consistent effects across four populations. Husbands' contribution to housework also has favorable influence in Taiwan and Japan. This study concludes that husbands and wives are more willing to talk their feelings, emotions or troubles out when they are practicing equality on household behaviors. Assortative pairing on demographic background predicts less variation of intimacy among couples.

Keyword Marital satisfaction · Assortative marriage · Household division of labor · Husband-wife relationship

1 Intimacy, Equity and Gender Ideology Among Married People in East Asia

In the tradition of East Asian familial ethics, the stressing of the importance of distinction, or status hierarchy loomed so large that, allegedly, it undermined romantic love or intimacy in conjugal relationships. It was widely believed that the prototypical bonds in human relationships, the most fundamental of which were father-son, ruler-minster/subject, and husband-wife, had to follow a sequencing principle. In other words, the precedence of husband over wife governed the relationships of a couple, the primary concern of which was not an individual's well-being, but the structural outcomes of social order and stability produced by following rigidly prescribed

M.-C. Tsai (✉)
Research Center for Humanities and Social Sciences, Academia Sinica, Taipei, Taiwan
e-mail: mtsai304@gate.sinica.edu.tw

M.-C. Tsai and N. Iwai (eds.), *Quality of Life in Japan*, Quality of Life in Asia 13,
https://doi.org/10.1007/978-981-13-8910-8_8

rules of conduct (Tu 1998). The Japanese conception of household (*ie*) appears to represent a dramatic case in which "the whole is larger than the combination of its parts." The Japanese household assumed a transcendental status in which the familial roles and obligations largely determined interpersonal relationships and the well-being of family members. The life courses of many men and women were "programmed in advance", many of them featuring absence of personal feeling or even active suppression of individual preferences and choices (Lebra 1998: 214). Slote (1998) describes the Confucian patriarchy's male dominance as breeding fear in the subordinated, that is, wife and children, toward the authority of father. The father represented a distant, authoritative figure, and they were "frequently abusive and demeaning toward wives, diminishing the women's spirit and the vitality and spontaneity of the relationship". Consequently, "most marriages remained distant, formal and frequently hostile; both parties withdrawn, unfulfilled, and despairing of ever finding the emotional sustenance that every human requires" (p. 42). In a psychoanalytical viewpoint, Slote (1998) proposed that the frustrated mothers turned to the children, particularly the sons, for the comfort and devotion that they (as well as their children) were not able to find from the husband (fathers), inciting a classical Oedipal situation (Slote 1998).

Despite recurrent reference to, and perhaps more often, critique of this Confucian ideology of husband-wife relationships and the patriarchy established alongside when examining modern East Asian families, social scientists are aware of its anachronism and have explored new types of family which can house more intimate relationships and emotional closeness between genders (Kaku 2013). More interestingly, empirical evidence gathered from the field indicates that conventional Confucian ethics has been surprisingly robust in influencing many aspects of family life such as gender ideology, filial values, patterns of coresidence, intergenerational exchanges, etc. (Raymo et al. 2015; Tsai and Yang 2017; Yeh et al. 2013; Yi and Chen 2014). With these research issues in mind, these scholars depict the family as a "public institution", detailing its structural and institutional configurations as well as its functions for society at large, and have understandably concluded that more continuities than gaps have been observed along with societal modernization. Yet rarely is the family examined as a "private institution", in which family members seek and practice intimate relationships. Empirical literature on the intimacy of Asian husbands and wives unfortunately has been lacking.

To fill the gap, this paper endeavors to examine personal intimacy by way of comparative investigation of the confiding behaviors of spouses in four East Asian societies. Unlike most recent studies that adopt measures of marital satisfaction or reported stresses to assess close relationships between spouses in the region (Chang 2013; Bessey 2015; Li et al. 2016; Oshio et al. 2013; Chang 2013; Qian and Sayer 2016), this study takes a more direct approach to observe marital quality by way of observing intimate relationships, that is, a sharing sense of self-disclosure or a willingness to reveal inner feelings and identity between husband and wife (Reis and Shaver 1988; Jamieson 1998, 1999). Intimacy merits closer examination particularly in the Confucian core societies, because conventional stereotypes of a mute husband and a denigrated wife continue to be used to indicate a dramatic lack of conjugal attachment in modern East Asian families. This study offers empirical evidence to evaluate this stereotypical perception. This constitutes the first contribution of this article. This study additionally assesses major factors in explaining how intimacy can evolve or is suppressed. Two competing arguments from assortative marriage and equality theories are tested in particular. The third contribution is the adoption of a cross-national design which examines congruence and differences among four Confucian societies, instead of relying on the findings from single countries, to make generalizable conclusions. Indeed, there exist increasing complaints that what is known about intimacy in modern society has been narrowly deduced from non-academic sources (Smart and Shipman 2004), and lack of empirical evidence has prevented a fuller understanding of its significance for personal relationships, especially in societies where it has conventionally been believed to be in short supply.

2 Theoretical Arguments

Intimate relationships have long been ignored in studying transitions of families in East Asia, as researchers are more interested in observing the distribution of power and authority between husband and wife, which presumably prevents intimate disclosure between couples (Kaku 2013; Raymo et al. 2015; Yi and Chang 2012). However, as modernization has advanced and income increased in the past decades in East Asia, there have also evolved significant changes in family structure in the direction of legitimating individual choice, interests, and desires, away from familial commitment, duties, and obligations. The postmodern individualization thesis (Beck and Beck-Gernsheim 1995; Bauman 2003) contends that families are shifting from a public institution to a private institution. Increasingly, people form unions for no other reason than an appreciation of each other's unique qualities. This value position has stirred up another round of moral ranting that post-modern marriage is more about finding oneself, alternatively, a personal journey of self-actualization. Marriage, for many contemporary people, is a binding file into which a series of passages and narratives of stories of intimacy along the life course are collected overtime. Each page tells a distinct tale. Whether the whole constitutes a meaningful structure is not an issue, however.

For empirical researchers the fundamental question is: what are the driving forces in modern society to realize this desired intimacy? There are two major arguments. The first one stresses the similarity of traits between couples. Assortative marriage facilitates intimacy. The second emphasizes equality by proposing that intimacy is not likely to occur when couples are unequal in resources and power. Equal footing forecasts intimacy. This section elaborates the two arguments and proposes relevant hypotheses.

The assortative marriage perspective contends that mating selection is far from random, but is very structured, such that family members become similar to each other in many major characteristics, such as personality, cognitive ability, attractiveness, etc. Because mating involves what has been called "intersexual selection", that is, the preference choice the members of one sex express for members of the opposite sex (Buss 1985), certain traits with high "market value" receive greater attention in mating. Education is a personal asset marked with a high value. Unsurprisingly, as a result of selection, couples become similar in terms of human capital as well as in other forms of socioeconomic status. The consequences of assortative marriage have been well documented. Societal inequality increases due to high prevalence of educational assortative marriage (Mare 1991; Blossfeld 2009; Schwartz 2013; Qian 2017). Homogamous couples also adopt a longer-term perspective in familial decision; for instance, they invest more in the household and are more likely to own a house (Wong 2016).

Gain in quality relationships for assortative marriage is less certain, despite some findings suggesting favorable outcomes. Similarity makes possible candid communication, cooperation and reciprocation among couples. Men and women are more likely to benefit from altruism, trust, and support in these well-intended interactions (Buss 1985; Reis and Shaver 1988). On the other hand, heterogamous unions may be associated with poorer relationship outcomes because of differences in lifestyles, values, and beliefs. Moreover, the partners' disparity in status, religious affiliation, or ethnic backgrounds is likely to incite disapproval from family members and kin (Schwartz 2013). In the American context, studies show that interracial dating and cohabiting couples are more likely to split up, and marriages across races and ethnicities are more likely to divorce (Blackwell and Lichter 2004; Joyner and Kao 2005; Zhang and Van Hook 2009). Similar outcomes are seen among the Asian population. In Taiwan, cross-border marriages with Southeast Asian people were found to have a high rate of divorce (Tsai and Chen 2017). However, there is a lack of direct testing of assortative marriage regarding its effect on intimacy. This issue is all the more intriguing against an increasing trend of homogamy in the East Asian context (Han 2010; Wu et al. 2013).

In contrast to the argument that similarity facilitates close relationships between partners, the intimacy perpective suggests the importance of equality between couples. Mutual disclosure and emotional closeness, which have been desired by modern couples, in fact have been prevented by gender inequality, rather than by categorical dissimilarity in status and resources. In modern families, men continue to play the role of a patriarchal father, which usually falls very short of fairness, mutual help and voluntary reciprocation. These favored features, ironically, are found more often in friendship. Jamieson (1998) proposes that only when couples become equals can intimacy evolve more easily and frequently. Giddens (1992) identifies lesbian and gay men as "pioneers" in this genre of "confluent love", highlighting a relationship of sexual and emotional equality (Roseneil and Budgeon 2004).

But the stories, narratives, and a revolution of intimacy observed from the "post-familial" partners should not be overdramatized. Many studies are based on highly selective samples and have limited generalizability (Roseneil and Budgeon 2004). A contribution of this approach should be highlighted, nevertheless. Desirable intimacy seems likely to be realized when a couple's relationship features equality and friendship, rather than being carefully matched in backgrounds. One partner needs to be treated as equal even if he or she lacks privileges or market value. Such equality is hard to come by for the reason that males, equipped with more power and influence, often fail to resist their masculinity and to relinquish their hold on benefits from inequality in the couple's relationship. Too easily the practice of intimacy is channeled into pre-established gendered scripts, which has been masked by heterosexual patriarchal structures (Jamieson 2011). Restated, intimacy is closely associated with equality between partners. When relationships are exercised on the basis of equal footing of the couple, pure relationships are more likely to happen, and confiding between husband and wife thus becomes a regular exchange in daily life.

Research on the East Asian family has not frequently investigated the intimacy of couples. Rather, perceived marital satisfaction receives more attention when it comes to understanding how couples evaluate their experiences about marriage and family. Xu and Lai (2004) investigated marital quality measured by two composite indicators, marital togetherness and marital satisfaction for men and women in Taiwan. They show a positive association of marital quality with equal marital power across sexes. Among married men, nontraditional division of labor increases marital quality, while this correlation is negligible among their counterparts. Taniguchi and Kaufman (2014) studied a national sample from Japan with a focus on both marital satisfaction and life satisfaction, and identified "relative income" (measured by comparing family income with the societal average), health and traditional values as strong predictors. They additionally show that men with traditional attitudes are more likely to do "emotional work", which seems to be contradictory to what is expected from pure relationship theory. Qian and Sayer (2016) extend their scope of observation by comparing four East Asian countries. They show interesting variation in which a greater share of housework in Japan and South Korea led to lower marital satisfaction for men only, while in Taiwan and China, this association is not observed. A possible explanation is that in the former two societies, men tend to be major income earners, while there also exists a stronger cultural belief in gendered division of labor.

In sum, this paper highlights two major hypotheses and tests them against a cross-national dataset in East Asia. In the following sections, I examine empirically whether assortative marriage and couple equality have substantial effects on the intimacy of men and women in this region. Particular attention to the measures of similarity and equality of the couples is exercised in order to obtain a fuller understanding of how these distinct configurations influence intimacy in the couples. To better capture the dyadic interaction between couples, an innovative approach is proposed for measurement of intimacy. Rather than treating the close relationships of husband and wife as separate phenomena, I stress that they tend to appear simultaneously, and that this has substantial implications for the choice of estimation techniques.

3 Data and Measurement

Data for this study are drawn from the 2006 East Asian Social Survey (EASS) (www.eassda.org). The EASS is a biennial social survey project consisting of four general social surveys in East Asia: the Chinese General Social Survey, Japanese General Social Survey, Korean General Social Survey, and Taiwan Social Change Survey. Currently there are four waves of the EASS survey, 2006, 2008, 2010, and 2012, which offer public access to the data. Each wave has a different theme. The theme of the 2006 EASS is family dynamics, with a specific focus on respondents' familial behaviors, values, and attitudes, which makes the 2006 EASS, despite being somewhat dated, a unique and appropriate dataset for our study on intimacy and its association with covariates. Representative adults aged 18 or older in China, Korea, and Taiwan and aged 20–89 in Japan were selected through multi-stage stratified random sampling. All surveys were conducted in face-to-face interviews in all four societies except for Japan, where some data were collected from self-administered questionnaires. The response rates varied by society: 38.5% in China with 3,208 respondents, 53.3% or 58.9% in Japan with 2,130 respondents (in-person vs. self-administered, respectively), 65.7% in South Korea with 1,605 respondents, and 42% in Taiwan with 2,102 respondents (ICPSR 2013). My analysis includes only those currently married, with an upper age limit of less than 70.

4 Variables

Confiding husband and wife. I conceptualize intimacy as willingness to disclose oneself to one's spouse. This approach is distinctive, as it emphasizes emotional exchange and close connection, in contrast to conventional approaches which capture it with an overall subjective evaluation of satisfaction with partners or with marital life. The latter is not satisfactory because it concentrates on self-reported rating without probing into mutual interactions between couples. The EASS provides the responses of married men and women to the question "my spouse listens to my troubles" and "my spouse speaks his/her troubles to me" (7-point scale, from "very true" = 7 to "not true at all" = 1). I recoded these two pieces of information to generate two dependent variables–husband confiding and wife confiding–by the following procedures.

Confiding husband is measured by the average score of:

- "my spouse listens to my troubles" for male respondents
- "my spouse speaks his/her troubles to me" for female respondents.

This operationalization generates an indicator of a husband that is willing to *talk* his troubles out. Similarly, a dependent variable of confiding wife is generated from:

- "my spouse speaks his/her troubles to me" for male respondents
- "my spouse listens to my troubles" for female respondents.

Two merits of this operationalization should be noted. It captures to some extent two-way communication rather than using single indicators to identify intimacy. That is, the dependent variable is observations from respondents who reported that they talked to spouses and their spouses listened. Another merit is that this measure addresses *verbal* sharing of encountered difficulties, perhaps even threats and vulnerabilities, with partners. It does not involve *expressive* sharing of one's inner emotions, as these might have been seen as feminine, "sissy stuff" which male respondents might deny or refuse to report outright in order to maintain their masculinity. Indeed, males tend to register a lower level of intimacy when measurement is more of emotional and expressive closeness than behavioral closeness (Patrick and Beckenbach 2009).

Assortative marriage. Two major dimensions of assortative marriage are considered. First, similarity in age is considered (see Tables 1 and 2 for summary statistics). Strictly, zero difference in age represents perfect matching. Yet in East Asia, as elsewhere, men tend to prefer having younger wives, not merely for greater sexual attraction, but also for expected care they can benefit from, as they get aged earlier and their health deteriorates. To fully capture the effect of age gradients, I designed a set of dummies of age difference by comparing the group of "no age difference" to "wife older than husband", "husband older by 1–3 years", "husband older by 4–6 years", "husband older by 7–9 years" and "husband older by 10 years or more". A recent study found that only when an age gap is 10 years or more do negative effects on marital satisfaction appear substantial (Lee and McKinnish 2018).

The second is educational assortative marriage. On the basis of educational degrees of both husband and wife, I generated a dummy design for comparing four groups: (1) both with college education (or more), (2) the husband has a college degree, but the wife does not, (3) the wife has a college degree but the husband does not; and (4) both husband and wife have no college degree. I expect the first group to show a higher level of intimacy, as these couples benefit not only from higher education but also similarity in educational background.

Inequality. Three measures of inequality between couples are adopted to reflect its diverse dimensions. The first one pertains to economic inequality of a couple when they get married. The wife's status in a "new" home can to some degree be determined by her original family's social status. This element is important because in many East Asian families, marriage has been believed to be a familial affair rather than merely a couple's decision for their private life. Indeed, in the past, clear status difference in origin family was disapproved of outright by parents. A marriage of "equal with equal" in terms of status and wealth is deemed appropriate. I expect that the better the economic status of a wife's family of origin is, the more likely her husband's equality and intimacy are to be enhanced as a consequence. I used the reported economic conditions and coded that a higher score indicates the husband's family had better economic conditions (=5, and 1 = wife's family have better conditions). This factor should be negatively correlated with the couple's confiding behavior.

The second measure taps economic contribution to the family by the husband and wife. Ideally, relative income can be used to represent different economic status of the husband and wife. However, the EASS did not collect information on the spouse's

Table 1 Percentage of category variables (%)

	Taiwan (n = 993)	South Korea (n = 899)	Japan (n = 598)	China (n = 2194)
Female (=1)	49.3	57.5	63.7	54.3
Education of husband				
Junior high or less	32.6	12.0	9.2	63.6
Higher secondary	32.2	35.0	50.2	24.3
Above higher secondary	35.3	52.0	40.6	12.2
Education of wife				
Junior high or less	38.9	17.4	7.7	71.9
Higher secondary completed	34.0	44.1	58.0	20.7
Above higher secondary	27.1	38.5	34.3	7.4
How the couple first met				
Arrangement	12.1	26.2	14.4	12.6
Introduction by other	40.2	34.1	29.9	59.1
By self	47.7	39.8	55.7	28.2
Age gap between couple				
Wife's older than husband	8.1	6.5	13.5	11.8
Same age	10.8	13.0	12.2	14.9
Husband's older: 1–3 years	38.7	40.0	38.6	44.0
Husband's older: 4–6 years	24.7	29.9	24.5	20.3
Husband's older: 7–9 years	9.4	8.4	8.1	6.3
Husband's older: ≥10 years	8.4	2.2	3.2	2.7
Educational difference between couple				
Both college	22.6	34.6	25.9	4.8
Husband college, wife no	12.7	18.4	14.7	7.4
Wife college, husband no	4.5	3.9	8.4	2.6
Both less than college	60.2	43.2	51.0	85.2
Employment matching between couple				
Both employed	60.4	44.1	75.9	69.0
Husband employed	24.5	48.1	21.7	19.8
Wife employed	5.8	3.5	2.3	9.4
Both not employed	9.4	4.3	0.0	1.8

Table 2 Summary of continuous variables

	Taiwan (n = 993)		South Korea (n = 899)		Japan (n = 598)		China (n = 2194)	
	Mean	S.D.	Mean	S.D.	Mean	S.D.	Mean	S.D.
Husband's age	47.0	10.3	44.9	9.3	49.1	10.4	45.1	11.4
Wife's age	43.6	10.2	42.0	8.9	46.6	9.8	42.9	11.1
Parent(s) influence on decision on marriage	2.1	1.0	2.4	1.0	2.0	0.9	2.2	1.0
Gender ideology	4.3	1.0	4.4	1.1	4.5	0.9	4.0	0.7
Pro-divorce attitude	4.0	1.8	3.7	1.8	4.1	1.1	4.3	1.3
Husband's economic condition better when married	3.0	0.9	–	–	–	–	3.1	0.7
Husband should be older than wife	4.9	1.5	4.5	1.4	3.8	1.0	4.6	1.1
Age gap between couple	3.4	3.8	2.9	2.8	2.4	3.5	2.2	3.0
Equal power in decision making	0.5	0.5	0.6	0.5	0.6	0.5	0.6	0.5
Housework by husband	1.4	1.7	1.1	1.6	0.6	1.1	2.3	2.3
Housework by wife	4.9	2.1	5.8	1.7	5.8	1.4	5.4	1.9

income. I thus used the employment status as a proxy. Respondents are regrouped into four categories: (1) both husband and wife are employed, (2) only husband is employed, (3) only wife is employed, and (4) both are unemployed. It is assumed that couples in the first category are more equal in economic status and thus are more intimate, although the limitation of this measurement is acknowledged.

Third, the share of decision-making power and division of labor in housework is particularly suitable for measurement of equality, as these indicators measure whether men and women share power in deciding on crucial household issues. Equal power in decision-making is measured by whether respondents report "common decision" (coded 1) in "children's discipline and education", "support for my own parents", "allocation of living expenses" and "purchase higher-cost items", respectively (0 = husband or wife solely decided). An average score was calculated for this predictor.

With regard to housework sharing, the EASS provides information on how often the respondents and their spouses prepared evening meals, did the laundry, and cleaned the house. I coded the measures in a way that shows the mean of the frequency with which a respondent did them in a week. This basically shows the *range* of housework done. As this is not a measure of the amount of time spent on housework, I am not able to calculate the proportion of housework each partner had done.

A number of controls are considered in this research design. Besides basic demographic background including gender (female = 1, male = 0), husband's and wife's age and education (college or more, secondary high school, and junior high school or less, with the latter used as reference), I also explore the influence of the way the couple first met (by themselves, introduced by others, or arrangement on purpose by parents, relatives, etc., with the latter being the reference group), and a measure of parental influence on their getting married (4 = very much, 1 = not at all or "parents were not alive at that time").

In addition, the factor of preference for age difference, tapped by response to the question "Husband should be older than wife" (scale same as the gender ideology measure), is also utilized in estimation. The motivation for this inclusion is that it allows examination of a belief about age difference, in contrast to objective age gradation as was mentioned previously. It is expected that agreement to this question reflects traditionalism and is thus negatively correlated with confiding behaviors.

Gender ideology has been regularly surveyed in marital quality research (Greenstein 1996; Wilcox and Nork 2006; Davis and Greenstein 2009; Qian and Sayer 2016). In this study, conventional gender ideology is measured by four questions:

- It is more important for a wife to help her husband's career than to pursue her own career.
- A husband's job is to earn money; a wife's job is to look after the home and family.
- Men ought to do a larger share of household work than they do now (reverse coding).
- During an economic recession, it is all right for women to be laid off prior to men.

These four questions are scored on 7-point scale, with "strongly agree" = 7 to "strongly disagree" = 1. They cover key issues concerning the prescribed roles for the husband and wife from a male bread winner model. Measured on a 7-point scale, from "strongly agree" = 7 to "strongly disagree" = 1). The level of correlation is quite satisfactory, with the internal consistency coefficient Cronbach's $\alpha = 0.64$ for China, 0.58 for Taiwan, 0.65 for South Korea and 0.71 for Japan. It is noted that Taiwan's lower α value indicates less agreement across the four questions. I expect that a stronger gender ideology decreases the disposition to self-disclose to one's partner (see Tables 1 and 2 for summary statistics).

Pro-divorce attitude. In addition to gender ideology, which represents ideas about the division of functional roles for the husband and wife, I further explore an additional value position regarding divorce as a potential explanatory factor. Modern families increasingly risk a choice of "exit" when a relationship does not work out. I use the following question to tap an *un*conventional familial attitude: "Divorce is best when marriage can't work out" (7-point scale). A higher score of pro-divorce attitude is expected to be associated with more frequently confiding.

Estimation. This study estimates the determinants of both of confiding husband and wife. There is high correlation between the two dependent variables: a husband who reported that his spouse listened to him is likely to say that his spouse also talked to him about her troubles. Thus, men, as well as women, tend to report both reciprocal interactions between couples. In this situation, application of single equation estimation can be inefficient and methodologically problematic, because the error terms across different regression equations are highly correlated. It is suggested that the information of correlated residuals be employed to improve the regression equations under consideration. The seemingly uncorrelated regression (SUR) method provides more efficient estimation in this situation than does the ordinary least squares method (even though the latter's estimation remains unbiased and consistent) (Felmlee and Hargens 1988; Zellner 1962). Note that when the cross-equation correlation is low or approaching zero, gain from this generalized least squares method can be trivial. Following the joint estimation approach, I perform the SUR estimation for both equations of confiding husband and wife simultaneously for each country.

5 Results and Analysis

Figure 1 presents the relative incidence of confiding couples across the four analyzed societies. I arrange the first three countries according to their relative income level; China is placed in the last column, because of its being a socialist country. Compared to the first three societies, China's national income and gender equality also lag far behind (United Nations Development Programme 2018). On a seven-point scale, confiding is reported more frequently in Taiwan and South Korea, especially among married males. On a seven-point scale, the average for the confiding Taiwan husband reaches 5.5, indicating that disclosure is not uncommon. The average score of confiding wife is somewhat lower but the difference between gender is trivial in magnitude. South Korea males register a substantially higher score of the husband being willing to talk. However, the mean for confiding wife is relatively low, with a marked difference of nearly 0.5 to the confiding husband. Although Japanese husbands have

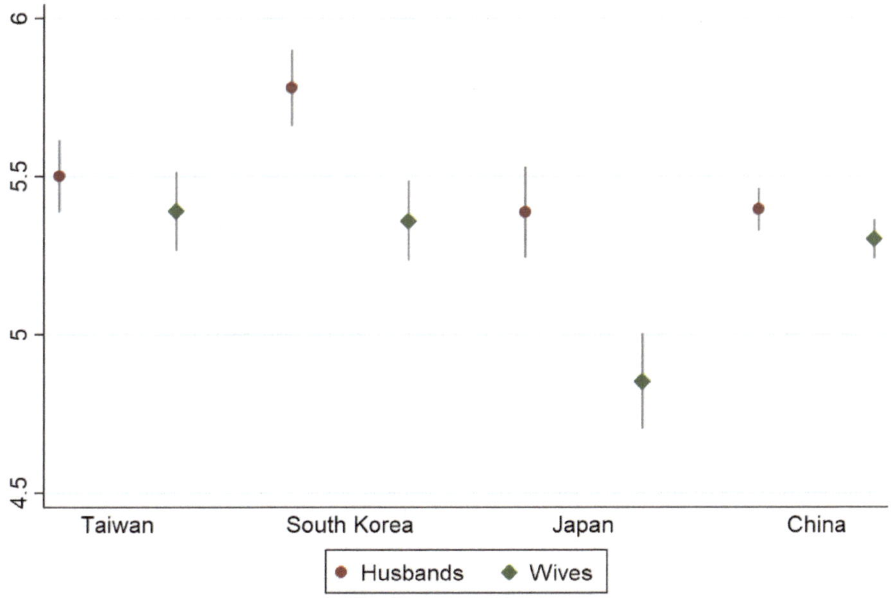

Fig. 1 Confiding husbands and wives in East Asia

been believed to be very reserved in emotional expression, the result is "counterintuitive", because the mean of confiding husband remains close to that of Taiwan. Yet, surprisingly, the score for confiding wife drops to a much lower level (approximately 4.8). This is partly because in our measurement married women reported that their spouses seldom listened to their act of confiding. There may be a vicious cycle here in which men listen reluctantly and therefore women confide infrequently. Among the Chinese, the variable of confiding husband has a score similar to that of Japan, and the score for confiding wife is close to those in Taiwan and South Korea.

Table 3 reports the SUR regression results across East Asia. The residuals' correlations across equations are substantial (ranging from 0.57 to 0.80, all $p < 0.001$), which justifies the use of a generalized least squares estimation. In Taiwan, as in other three countries, females tend to report that their husbands confide less frequently. Gender difference in intimate disclosure as such has been documented in the Western societies (Jamieson 1998). Female respondents, however, do not report that wives confide less often in Taiwan, South Korea and China (see the first row of the table). Yet Japanese women express an opinion that "wife confiding" happens infrequently. It is not suitable to conclude herein that Japanese women are *solely* responsible for the lower mutual communication. Holding emotions within oneself seems to be normative for Japanese husbands, such that wives find less opportunities to talk about their troubles to their partners. This might result in less confiding by wives, as was reported by Japanese women.

In the second row, it is shown that older husbands in Taiwan confided less than younger ones. Note that in this equation, wife's age is not included to avoid high collinearity with "age matching" between couples. One additional reason is that the SUR estimation reduces to the OLS result when the equations in the system carry identical predictors (Felmlee and Hargens 1988). The findings demonstrate that for older wives' score on confiding to their husband decreases among Taiwanese and Chinese. Judging from the sizes of the regression coefficients, gender difference is more marked in South Korea and Japan, while age difference deters confiding more strongly in Taiwan and China.

Do husbands with better education confide more frequently? My finding across countries does not provide supportive evidence. However, in Taiwan and South Korea, husband's education indeed increases the likelihood of wife confiding. Wives' confiding is not related with their own education across the four societies. In general, education is not a strong predictor in estimation.

How the couple first met has no influence on confiding of either men or women. Perhaps this should not be surprising, as for many respondents, how they dated with their spouses a long time before carries no weight in how they confide nowadays. Even if they met by themselves, which indicates less family intervention in spousal choice, such autonomy is a very distant factor for evaluating their current relationships. In comparison, parental influence in the marriage decision appears more important but the effect is mixed. It helps facilitate confiding of the husbands in China, but not elsewhere. It reduces confiding of wives in Taiwan but it has positive effects in South Korea and China. In the Asian context, parental influence has been a screening mechanism for finding a "perfect" match for adult children. For

Table 3 Confiding of husbands and wives in East Asia (seemingly unrelated regression estimates)

Dependent variables: confiding	Taiwan		South Korea		Japan		China	
	Husband	Wife	Husband	Wife	Husband	Wife	Husband	Wife
Female (=1)	−0.26** (0.09)	−0.13 (0.09)	−0.80*** (0.09)	−0.05 (0.10)	−0.48*** (0.12)	−0.45*** (0.13)	−0.11* (0.05)	−0.08 (0.05)
Husband's age	−0.01* (0.01)		0.01 (0.01)		0.01 (0.01)		−0.01*** (0.00)	
Wife's age		−0.02*** (0.01)		0.00 (0.01)		−0.00 (0.01)		−0.01** (0.00)
Education of husband (ref = junior high on less)								
Higher secondary	0.10 (0.12)	0.09 (0.12)	0.24 (0.19)	0.02 (0.20)	−0.08 (0.20)	0.13 (0.21)	0.09 (0.06)	0.15* (0.06)
Above higher secondary	0.13 (0.14)	0.32* (0.14)	0.36 (0.21)	0.44* (0.22)	−0.05 (0.22)	0.27 (0.23)	0.07 (0.09)	0.11 (0.09)
Education of wife (ref = junior high on less)								
Higher secondary completed	0.11 (0.12)	−0.02 (0.12)	−0.31 (0.18)	−0.11 (0.18)	−0.31 (0.21)	−0.38 (0.23)	−0.00 (0.06)	−0.05 (0.06)
Above higher secondary	0.21 (0.16)	0.04 (0.15)	−0.06 (0.22)	0.15 (0.22)	0.03 (0.24)	−0.25 (0.25)	0.21 (0.11)	0.16 (0.10)
How couple first met (ref = arrangement)								
Introduction	−0.05 (0.14)	−0.12 (0.14)	−0.02 (0.13)	0.13 (0.13)	0.12 (0.17)	0.30 (0.18)	0.03 (0.07)	0.11 (0.07)

(continued)

Table 3 (continued)

Dependent variables: confiding	Taiwan		South Korea		Japan		China	
	Husband	Wife	Husband	Wife	Husband	Wife	Husband	Wife
By themselves	0.14 (0.14)	−0.06 (0.14)	0.06 (0.13)	0.09 (0.13)	−0.02 (0.16)	0.14 (0.17)	0.04 (0.08)	0.14 (0.08)
Parent(s) influenced marriage decision	−0.05 (0.04)	−0.10* (0.04)	0.02 (0.05)	0.11* (0.05)	−0.05 (0.06)	−0.03 (0.06)	0.07** (0.02)	0.07** (0.02)
Husband should be older than wife	0.07* (0.03)	0.05 (0.03)	0.10** (0.04)	0.10** (0.04)	−0.07 (0.05)	−0.06 (0.06)	0.05* (0.02)	0.03 (0.02)
Conventional gender ideology	0.01 (0.05)	0.03 (0.05)	0.06 (0.05)	−0.07 (0.05)	−0.10 (0.07)	−0.12 (0.07)	−0.03 (0.03)	−0.01 (0.03)
Pro-divorce attitude	0.01 (0.02)	−0.04 (0.02)	−0.03 (0.03)	−0.00 (0.03)	−0.01 (0.05)	−0.05 (0.05)	0.05** (0.02)	0.05** (0.02)
Age gap between couple (ref = same ages)								
Wives younger than husbands	0.23 (0.19)	0.17 (0.19)	0.12 (0.21)	−0.02 (0.21)	0.05 (0.19)	−0.05 (0.21)	−0.09 (0.09)	−0.02 (0.09)
Husband's older: 1−3 years	−0.06 (0.14)	−0.04 (0.14)	0.09 (0.14)	0.03 (0.15)	0.19 (0.17)	0.07 (0.18)	−0.06 (0.07)	−0.05 (0.07)

(continued)

Table 3 (continued)

Dependent variables: confiding	Taiwan		South Korea		Japan		China	
	Husband	Wife	Husband	Wife	Husband	Wife	Husband	Wife
Husband's older: 4–6 years	0.09 (0.15)	−0.09 (0.15)	−0.18 (0.15)	−0.24 (0.15)	0.00 (0.18)	−0.07 (0.19)	−0.04 (0.08)	−0.08 (0.08)
Husband's older: 7–9 years	0.32 (0.19)	0.09 (0.18)	−0.20 (0.21)	−0.01 (0.21)	−0.24 (0.24)	−0.25 (0.25)	0.03 (0.12)	−0.05 (0.12)
Husband's older: more than 10 years	0.01 (0.22)	−0.72*** (0.22)	0.26 (0.36)	0.77* (0.36)	0.34 (0.35)	−0.18 (0.37)	−0.09 (0.16)	−0.28 (0.16)
Husband's family is better in economic situation	0.00 (0.05)	0.05 (0.05)					−0.08* (0.04)	−0.01 (0.04)
Employment status (ref = both employment)								
Husband employed	0.03 (0.10)	−0.18 (0.10)	−0.16 (0.10)	−0.09 (0.10)	−0.28* (0.14)	−0.04 (0.14)	0.13* (0.06)	0.15* (0.06)
Wife employed	0.02 (0.19)	−0.28 (0.19)	−0.11 (0.26)	−0.47 (0.26)	0.27 (0.35)	0.36 (0.37)	0.06 (0.09)	0.05 (0.09)
Both no employed	0.02 (0.17)	0.14 (0.16)	0.04 (0.23)	−0.12 (0.24)			0.22 (0.18)	0.23 (0.18)

(continued)

Table 3 (continued)

Dependent variables: confiding	Taiwan		South Korea		Japan		China	
	Husband	Wife	Husband	Wife	Husband	Wife	Husband	Wife
Shared decision-making	0.30*** (0.09)	0.34*** (0.08)	0.38*** (0.09)	0.43*** (0.09)	0.41*** (0.10)	0.49*** (0.11)	0.12* (0.05)	0.12** (0.05)
Housework by husband		0.05** (0.02)		0.07** (0.02)		0.10** (0.04)		−0.01 (0.01)
Housework by wife	0.05** (0.02)		0.01 (0.02)		−0.01 (0.03)		0.01 (0.01)	
Constant	5.18*** (0.48)	6.03*** (0.46)	4.53*** (0.57)	4.42*** (0.56)	5.83*** (0.66)	6.23*** (0.67)	5.37*** (0.26)	5.08*** (0.26)
R^2	0.06	0.10	0.12	0.10	0.12	0.11	0.03	0.03
Correlation of residuals	0.63***		0.57***		0.63***		0.80***	
Breusch−Pagan test	398.4***		293.6***		235.3***		1393.9***	
Observations	993		899		598		2,194	

Standard errors in parentheses; $^{***}p < 0.001$, $^{**}p < 0.01$, $^{*}p < 0.05$

parents who played an active role in this regard, the outcomes are not identical across societies. The unfavorable results for the confiding wife in Taiwan might reflect this society's stronger culture of feminism that demands more freedom for female adults. Probably in such a context, overly strong involvement by parents in partner choice incurs a negative effect for wives' communication with husbands.

Three potential attitudes are tested for their association with confiding. Respondents who believe that the husband should be older than the wife tend to confide more frequently in Taiwan, South Korea, and China. In South Korea, this belief also leads to a higher likelihood of reporting a confiding wife. These results are not what I had expected. It is indeed a puzzle for future researchers.

Conventional gender ideology shows little influence on confiding of husbands or wives. Pro-divorce attitude, a proxy of value modernization at the individual level, is not associated with confiding of both husband and wife in East Asian societies except China, where a positive correlation is noted. It seems that along a massive transition toward the market economy, Chinese couples have become, in a sense, progressive, in that intimacy is more sought after for those who become modernized in their value position.

We now come to the testing of assortative marriage. Table 3 does not include educational assortative marriage, because this aspect of marriage matching is not very powerful in explanation. Only in China, when both partners have college education or more, are they more likely to report a confiding husband. This pattern is also observed for confiding wife (not shown to save space). However, in other three countries there was no linkage between assortative marriage in terms of education and the likelihood of confiding. Educational assortative marriage fails to predict confiding in Taiwan, South Korea and Japan. Therefore, I decided not to take it further into the following analysis (Fig. 2).

The next type of assortative marriage is concerned with similarity in age. As is shown from equations for the four samples, age gradients as were operationalized

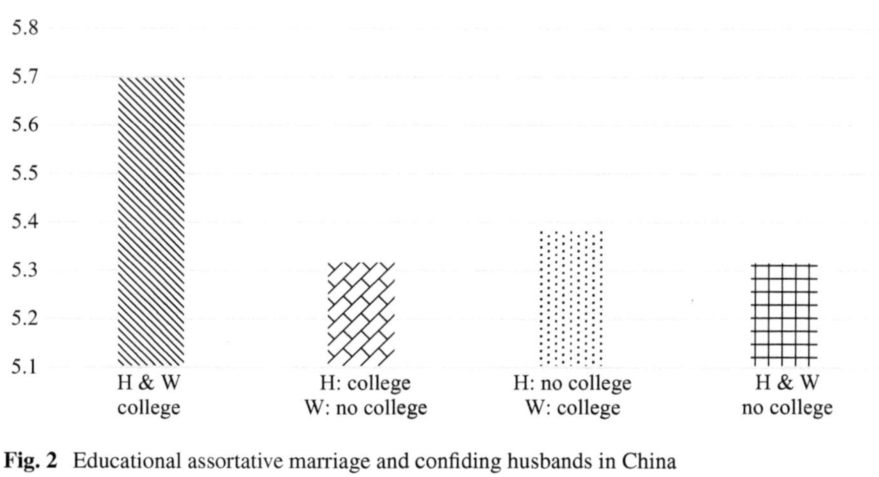

Fig. 2 Educational assortative marriage and confiding husbands in China

here are not effective in prediction. In Taiwan, age difference larger than 10 years carries a negative impact for the confiding wife variable. Yet, in South Korea, this large age gap produces a favorable influence. Overall, age gradation is not a crucial factor; when it exerts effects in different societies, their results are more mixed than consistent.

I now move to examine the equality-intimacy hypothesis. A plethora of equality measures are mobilized for testing. The first one is economic status of the origin family. The EASS provides this information only for Taiwan and China. In Taiwan, this factor is not correlated with confiding. In China, the better the economic status of the husband's family, the less he talks about his troubles. That means that in a marriage in which wives are comparable to husbands in economic standing, their husbands are more likely to confide. This piece of evidence can be seen as supportive. However, the data are available only in Taiwan and China, so the finding may not be generalizable over East Asia.

Employment is used as a proxy of income contribution to the household in this study. The group of both couples having a job is designated as a reference group, to which other types of employment are contrasted. I found that in Japan, when only the husband is employed, his confiding declines significantly, an outcome in favor of the equality hypothesis. In China, the opposite is obtained, and the significant influence is observed in the equations of both confiding husband and wife. In Taiwan and South Korea, no substantial effects are detected. Thus, the outcomes are mixed at best.

Next, shared decision-making is examined. Across the four East Asian countries, equal power in making decisions on parenting or household consumption facilitates confiding for husbands and wives in a consistent way. This is the most favorable evidence for a hypothesis proposed in this study.

Contribution to household work by husbands, a proxy of gender equality, enhances confiding of wives in Taiwan and Japan. The coefficient in the equation for the Japanese is quite high ($b = 0.10$, $p < 0.01$), twice that in Taiwan ($b = 0.05$, $p < 0.05$). Interestingly, the wife's increased housework also facilitates confiding from husbands in Taiwan, but not in other samples. This seems to be an unusual sort of exchange between husbands doing emotional exchange and wives doing housework. Equations for the other three societies do not show similar substantial results.

6 Discussion and Conclusion

Emotional disclosure, a key element in defining intimate relationships between couples in modern society, has been investigated infrequently in East Asia. Conventional scholarship provides insufficient evidence on this crucial aspect of personal life for Asian couples, such that to what extent intimate connections have been firmly built can only be gauged indirectly using a vague, though widely used, indicator of "marital satisfaction". In responding to the current theories that propose to examine the patterns of intimacy more systematically (Jamieson 1998), this study endeavors to analyze survey data from four East Asian countries and provides several important findings for cross-national comparison. First, confiding by husbands and wives is not uncommon, a result in contrast to what has been described in past literature, namely that disclosure has been constrained in patriarchal relationships (Kaku 2013). It can be safely asserted that intimacy is desired and sought after by men and women in East Asian families. Second, females have indeed complained about their husbands' reluctance in emotional disclosure. Across the four societies, females consistently reported a lower score of confiding by husbands as well as themselves. This gender gap is unsurprising. Men tend to overstate their emotional closeness with their partners, just as they overestimate their contribution to housework and sharing of responsibilities (Bianchi et al. 2000; Shelton and John 1996). Third, in comparison to conventional ideology and the form of marriage arrangement, gender equality represented by power sharing in decision-making is the most consistent factor in explaining how intimacy can be enhanced. Housework sharing by the husband, another proxy of gender equality, also generates favorable influence on the wife confiding in Taiwan and Japan. The equal power—intimacy hypothesis is supported. Finally, the similarity factor which has been stressed in the assortative marriage approach is weak in prediction. Marriage matching in terms of educational background or age does not substantially facilitate intimacy. These findings provide important implications for reconsidering major interpretations of family and marriage in East Asia.

A fuller understanding of the changing East Asian family cannot be realized without recovering the role of intimacy in union formation and in sustaining relationships in this region. Intimate disclosure is desired even in a cultural context conventionally characterized by a Confucian patriarchy. As East Asia advances in income level and moves toward a post-industrial phase, personal life may become a central focus for younger generations. "Progressive" values such closeness, sharing, trust, and privacy, along with mutual understanding and self-autonomy, indeed have been popular in daily discourses on marriage and personal relationships. Unequal power, dominance and condescending attitudes increasingly are not justified or tolerated. Against this background, willingness to share power and resources on an equal footing appears to have strong implications for how private life in family can be satisfying. I stress that this thesis is very generalizable and is not necessarily constrained by a distinct Confucian culture which has been perceived to suppress intimate relationships between married couples in East Asia.

Similarities (rather than opposites) are said to be a solid foundation for marriage. However, this study provides little evidence for this assortative marriage argument. Homogamy is not a sufficient factor for intimate disclosure. It perhaps serves as a necessary condition at best, a setting in which conjugal interactions evolve but the relationship outcomes can vary across couples. In an unusual study of Finnish families, couples with higher education were reported to be more unfaithful. Yet, the fact that their lower-educated counterparts had fewer extramarital sexual affairs might reflect a lack of opportunities than a strong motivation of fidelity (Kontula and Haavio-Mannila 2004). Similarity in status and resources of the couples shows that marriage matching may be a well-planned, rational choice, but it does not guarantee intimacy.

In accordance with intimacy theory, the experience of mutual disclosure for a couple is more likely to appear when their relationship features equal power. This study conceptualizes equality as the practicing of common decision-making in major domains, instead of enforcing the preference of either person over the other. Equality as such can be crafted even if the partners are dissimilar in their origins, age, educational achievement, or income contribution. Equality as I measure it represents togetherness in fulfilling household chores or responsibilities, the accomplishment of which for most people is quite uninteresting or unpleasant. But it is through this willingness in doing daily chores or tasks that mutual exchange of ideas, thoughts, life goals, and so on is engendered or encouraged. When exchanges as such subside, the relationship between a couple moves close to an "empty-shell marriage", in which reciprocity and intimacy are hard to come by.

A number of research limitations should be noted. First, my measures of confiding behavior are not optimal. Ideally, a measure of mutual disclosure should include observations from interactions of a dyad in reciprocation or mutual emotional exchange. The EASS data provided relevant information gathered separately from each husband and wife, rather than a couple jointly telling what they had done about mutual disclosure. A research design which solicits information from both partners in a joint interview for richer, detailed reciprocal exchanges would be more helpful for untangling the intricate phenomenon of intimacy in a couple. Second, not all major aspects of homogamy are included. In the case of Taiwan, ethnic background has been one salient feature in assortative marriage (Hsieh and Chen 2012). Urban-rural difference and migration status should be key factors in China and South Korea. Regional origin may have some implications in Japan. Future research might incorporate these additional characteristics into the research design. Last, admittedly, the EASS data of 2006 this study uses are somewhat outdated. Fortunately, the EASS continues to operate in this region. When newer data are released for public use, a comparative study of intimate relationships and other familial behaviors in the analyzed four countries over time should be an interesting priority.

References

Bauman, Z. (2003). *Liquid love: On the frailty of human bonds*. Cambridge: Polity.

Beck, U., & Beck-Gernsheim, E. (1995). *The normal chaos of love*. Cambridge: Polity.

Bessey, D. (2015). Love actually? Dissecting the marriage–happiness relationship. *Asian Economic Review, 29*, 21–39. 513–530.

Bianchi, S. M., Milkie, M. A., Sayer, L. C., & Robinson, J. P. (2000). Is anyone doing the housework? Trends in the gender division of household labor. *Social Forces, 79*, 191–228.

Blackwell, D. L., & Lichter, D. T. (2004). Homogamy among dating, cohabiting, and married couples. *Sociological Quarterly, 45*, 719–737.

Blossfeld, H. (2009). Educational assortative marriage in comparative perspective. *Annual Review of Sociology, 35*, 513–530.

Buss, D. M. (1985). Human mate selection. *American Scientist, 73*(Jan-Feb), 47–51.

Chang, W.-c. (2013). Family ties, living arrangement, and marital satisfaction. *Journal of Happiness Studies, 14*(215–233), i.

Davis, S. N., & Greenstein, T. N. (2009). Gender ideology: Components, predictors, and consequences. *Annual Review of Sociology, 35,* 87–105.
Felmlee, H., & Hargens, L. L. (1988). Estimation and hypothesis testing for seemingly unrelated regression: A sociological application. *Social Science Research, 17,* 384–399.
Giddens, A. (1992). *The transformation of intimacy: Sexuality, love and eroticism in modern societies.* Cambridge: Polity.
Greenstein, T. N. (1996). Gender ideology and perceptions of the fairness of the division of household labor: Effects on marital quality. *Social Forces, 74,* 1029–1042.
Han, H. (2010). Trends in educational assortative marriage in China from 1970 to 2000. *Demographic Research, 22,* 733–770.
Hsieh, Y.-S., & Chen, I.-C. (2012). Changes in intergenerational influences on cross-ethnic marriage in Taiwan. In C.-C. Yi & Y.-H. Chang (Eds.), *Social change in Taiwan, 1985–2005: Family and marriage* (pp. 181–227). Insitute of Sociology, Academia Sinica: Taipei.
ICPSR. (2013). *East Asian social survey (Eass), cross-national survey data sets: Families in East Asia, 2006 (Icpsr 34606).* Retrieved March 15, 2017, from (http://www.icpsr.umich.edu/icpsrweb/ICPSR/studies/34606).
Jamieson, L. (2011). Intimacy as a concept: Explaining social change in the context of globalization or another form of ethnocentrism? *Sociological Research Online, 16*(4), 15. www.socresonline.org.uk/16/4/15.html.
Jamieson, L. (1998). *Intimacy: Personal Relationships in modern society.* Cambridge: Polity.
Jamieson, L. (1999). Intimacy transformed? A critical look at the 'pure relationship'. *Sociology, 33,* 477–494.
Joyner, K., & Kao, G. (2005). Interracial relationships and the transition to adulthood. *American Sociological Review, 70,* 563–581.
Kaku, S. (2013). *Patriarchy in East Asia: A comparative sociology of gender.* Leiden: Brill.
Kontula, O., & Haavio-Mannila, E. (2004). Renaissance of romanticism in the era of increasing individualism. In J. Jean Duncombe, K. Harrison, G. Allan, & D. Marsden (Eds.), *The state of affairs: Explorations in infidelity and commitment* (pp. 79–102). London: Routledge.
Lebra, T. S. (1998). Confucian gender order and personal fulfillment for Japanese women. In W. H. Slote & G. A. De Vos (Eds.), *Confucianism and the family* (pp. 209–227). Albany: State University of New York Press.
Lee, W.-S., & McKinnish, T. (2018). The marital satisfaction of differently aged couples. *Journal of Population Economics, 31,* 337–362.
Li, A., Robustelli, B. L., & Whisman, M. A. (2016). Marital adjustment and psychological distress in Japan. *Journal of Social and Personal Relationships, 33,* 855–866.
Mare, R. D. (1991). Five decades of educational assortative mating. *American Sociological Review, 56,* 15–32.
Oshio, T., Nozaki, K., & Kobayashi, M. (2013). Division of household labor and marital satisfaction in China, Japan, and Korea. *Journal of Family and Economic Issues, 34,* 211–223.
Patrick, S., & Beckenbach, J. (2009). Male perceptions of intimacy: a qualitative study. *Journal of Men's Studies, 17,* 47–56.
Qian, Y. (2017). Gender asymmetry in educational and income assortative marriage. *Journal of Marriage and Family, 79,* 318–336.
Qian, Y., & Sayer, L. C. (2016). Divison of labor, gender ideology, and marital satisfaction in East Asia. *Journal of Marriage and Family, 78,* 383–400.
Raymo, J. M., Park, H., Xie, Y., & Yeung, W. J. J. (2015). Marriage and family in East Asia: Continuity and change. *Annual Review of Sociology, 41,* 471–492.
Reis, H. T., & Shaver, P. (1988). Intimacy as an interpersonal process. In S. W. Duck (Ed.), *Handbook of personal relationships.* New York: Wiley.
Roseneil, S., & Budgeon, S. (2004). Cultures of intimacy and care beyond 'the family': Personal life and social change in the early 21st century. *Current Sociology, 52,* 135–159.
Schwartz, C. R. (2013). Trends and variation in assortative mating: Causes and consequences. *Annual Review of Sociology, 39,* 451–470.

Shelton, B. A., & John, D. (1996). The division of household labor. *Annual Review of Sociology, 22,* 299–322.
Slote, W. H. (1998). Psychocultural dynamics within the confucian family. In W. H. Slote & G. A. De Vos (Eds.), *Confucianism and the family* (pp. 37–51). Albany: State University of New York Press.
Smart, C., & Shipman, B. (2004). Vision in monochrome: Family, marriage and the individualization thesis. *British Journal of Sociology, 55,* 491–509.
Taniguchi, H., & Kaufman, G. (2014). Gender role attitudes, trouble talk, and marital satisfaction in Japan. *Journal of Social and Personal Relationships, 31,* 975–994.
Tsai, M.-C., & Chen, W.-C. (2017). *Family, work and wellbeing in Asia*. Dordrecht, The Netherlands: Springer.
Tsai, M.-C., & Yang, W.-S. (2017). Familial exchange and intergenerational contact in East Asian societies. In M.-C. Tsai & W.-C. Chen (Eds.), *Family, Work and Wellbeing in Asia* (pp. 21–42). Dordrecht: Springer.
Tu, W.-M. (1998). Probing the "three bonds" and "five relationships" in confucian humanism. In W. H. Slote & G. A. De Vos (Eds.), *Confucianism and the family* (pp. 121–136). Albany: State University of New York Press.
United Nations Development Programme. (2018). *Human development indices and indicators 2018 statistical update*. New York: United Nations.
Wilcox, W. B., & Nock, S. L. (2006). What's love got to do with it? equality, equity, commitment and women's marital quality. *Social Forces, 84,* 1321–1345.
Wong, H.-P. C. (2016). Ethnic assortative matching in marriage and family outcomes: Evidence from the mass migration to the US during 1900–1930. *Journal of Population Economics, 29,* 817–848.
Wu, L.-H., Yeh, H.-J., & Tsay, R. M. (2013). Meeting the other half: Matchmakers and settings of contact in educational assortative marriage (in Chinese). *Taiwanese Sociology, 26,* 147–190.
Xu, X., & Lai, S.-C. (2004). Gendern ideology, marital roles, and marital quality in Taiwan. *Journal of Family Issues, 25,* 318–355.
Yeh, K.-H., Yi, C.-C., Tsao, W.-C., & Wan, P.-S. (2013). Filial piety in contemporary Chinese societies: A comparative study of Taiwan, Hong Kong, and China. *International Sociology, 28*(3), 277–296.
Yi, C.-C., & Chang, Y.-H. (2012). *Social change in Taiwan, 1985–2005: Family and marriage*. Taipei: Insitute of Sociology, Academia Sinica.
Yi, C.-C., & Chen, Y.-H. (2014). The intergenerational transmission of the value of children in contemporary Chinese families: Taiwan and Mainland China compared. *Comparative Population Studies, 39*(4), 679–706.
Zellner, A. (1962). An efficient method of estimating seemingly unrelated regressions and tests for aggregation bias. *Journal of American Statistical Association, 57,* 348–368.
Zhang, Y., & Van Hook, J. (2009). Marital dissolution among interracial couples. *Journal of Marriage and Family, 71,* 95–107.

Ming-Chang Tsai is a Research Fellow and Deputy Director of Research Center for Humanities and Social Sciences, CHSS, Academia Sinica, Taiwan. He is President of the International Society for Quality of Life Studies, and was former president of Taiwanese Sociological Association and President of Research Committee of Social Indicators (RC55) of the International Sociological Association. His current research project focuses on family relationship, youth, and ageing in comparative perspective. His articles have appeared in *Comparative Sociology*, *Globalizations*, *Journal of Marriage and Family*, *Journal of Contemporary Asia*, *Sociological Research Online*. His recent books include *Global Exposure in East Asia* (Routledge, 2015) and *Family, Work and Wellbeing in Asia* (coeditor, Springer, 2017) (mtsai304@gate.sinica.edu.tw).

Quality of Life in Japan and Emigration: The Perspectives of Japanese Skilled Immigrants in Australia

N. Oishi and I. Hamada

Abstract The number of Japanese citizens living outside Japan has been increasing in the last four decades: it has risen almost threefold since the mid-1980s. What drives Japanese nationals to move away from their home country? Drawing on the narratives of 32 Japanese skilled immigrants in Australia, this study argues that their perceptions of QOL haves played a major role. These perceptions differ across individuals' life stage and gender, and particularly the time of immigration. This study found distinctive differences in the QOL perceptions between those who arrived before the Great East Japan Earthquake and Fukushima disasters in 2011 and those who came after 2011. Many studies have already pointed out the significance of "lifestyle migration," where non-economic factors have attracted Japanese immigrants to Australia. While acknowledging its importance for those who arrived before 2011, this study found that the most prominent drivers for the post-2011 Japanese immigrants were not lifestyle, but rather their acute perceptions of environmental, economic, and socio-political risks that could undermine the long-term QOL for themselves and their families.

Keywords Quality of life · QOL · Japanese skilled immigrants · Australia · Lifestyle migration · Work-life balance · Gender equality · Environmental risks · Disaster

1 Introduction: Growing Immigration from Japan to Australia

As the third-largest economy in the world, Japan has long been known for its economic prosperity. Recent assessments of its quality of life (QOL) are, however, rather mixed. On one hand, Japan is highly rated. For example, the Quality of Life Index

N. Oishi (✉) · I. Hamada
Monash University, Melbourne, Australia
e-mail: nana.oishi@unimelb.edu.au

M.-C. Tsai and N. Iwai (eds.), *Quality of Life in Japan*, Quality of Life in Asia 13,
https://doi.org/10.1007/978-981-13-8910-8_9

(QLI) 2017,[1] which is based on a large survey of global expatriates, placed Japan in the top tier: it ranked sixth among 65 countries in the world, and second in the QLI sub-index on "personal safety" (InterNations 2017). Japan also performed well in several measures of well-being in the Better Life Index[2] (OECD 2017a, b) . It ranked at the top in personal security and above the OECD average in income and wealth, education and skills, jobs and earnings, personal security, and environmental quality. The UNDP's Human Development Index (HDI), which is considered a "conventional gauge of living standards" (*Economist* 2000), placed Japan 17th among 188 countries in 2016 (UNDP 2016). While it is not as high as the ranks in other indices, Japan is still ranked in the 10% range of all countries in the world.

Despite Japan's generally high quality of life, the data show that emigration from Japan to other countries has been growing rapidly in the last two decades. According to data from the Japanese Ministry of Foreign Affairs (MOFA), which captures the stock of both long-term residents[3] and permanent residents, the total number of Japanese citizens staying overseas increased by 66.6%, from 811,712 in 2000 to 1,351,970 in 2017 (MOFA 2002, 2018). While this reflects a globalization of Japanese society and economy, resulting in an increase in temporary emigrants such as corporate expatriates and students, permanent emigration from Japan has also been increasing. The number of Japanese citizens who acquired permanent residency in other countries rose by an even higher rate (73.8%), from 278,619 in 2000 to 484,150 in 2017 (MOFA 2002, 2018). In particular, the pace of Japanese emigration has been accelerating since the Great East Japan Earthquake and the ensuing *tsunami* and Fukushima disasters in 2011. The number of Japanese permanent emigrants grew by 21.1% between 2011 and 2017 (MOFA 2018).

Overall, the increase in overseas Japanese citizens in 2011–2017 accounts for 12.5% of the country's net population loss during the same period. In other words, Japan's population decline is not simply caused by low birth rates, but also by the emigration of Japanese citizens, particularly since 2011. While some long-term residents of other countries may return to Japan eventually, others may stay permanently. Therefore, the increase in emigrants could have a significant impact on Japan's demographic future.

Australia has always been a popular destination for Japanese emigrants, but it has become even more popular since 2011. The number of Japanese citizens (long-term residents and permanent residents) living in Australia increased by 30.2%, from

[1]This index is based on the "Expat Insider" survey of "more than 12,500 respondents representing 166 nationalities and living in 188 countries or territories" who answered the questions about leisure options, travel and transport, health and well-being, safety, and security, as well as personal happiness (InterNations 2017).

[2]This is an aggregate index of 11 indices, including housing, income, jobs, education, life satisfaction and health (OECD 2017a, b).

[3]The category "long-term residents" encompasses a wide variety of Japanese citizens: some are the employees of Japanese corporations and their family members, while others are students, researchers, working holiday makers, and so on. Although this category is often perceived as temporary migrants, it actually includes prospective permanent migrants who are currently working on temporary skilled visas while waiting to receive permanent residency.

74,679 in 2011 to 97,223 in 2017 (MOFA 2012, 2018). The increase of Japanese permanent residents alone has been even higher—it rose by 32.9% during the same period. Australia is now the second most popular permanent destination for Japanese citizens after the United States. While the US still accommodates the largest number of Japanese citizens with permanent residency, the growth rate of Japanese permanent immigration to Australia for the last six years has been higher than that of the US.

Why is it that so many Japanese citizens have been leaving the country? Why do many of them choose Australia? In fact, the assessments of Australia's QOL are also mixed: Australia was ranked very high on the Human Development Index and the OECD's Better Life Index—second and third, respectively—but it was ranked 23rd on the Quality of Life Index, which is much lower than Japan (InterNations 2017; OECD 2017a, b; UNDP 2016) . In other words, the numerical indicators do not clearly indicate that Australia has a higher QOL than Japan. Therefore, to answer the question as to why many Japanese citizens are now leaving for overseas as permanent residents, especially to Australia, we need to investigate the qualitative and subjective perspectives of the Japanese immigrants themselves. How did they perceive their QOL in Japan and what kind of QOL dimensions were they most dissatisfied with? Did their QOL improve after immigrating to Australia? While existing studies have addressed the dimensions of lifestyle (Mizukami 2007; Nagatomo 2008, 2015; Sato 2001), no specific research has been conducted into the overall QOL in Japan from a broader perspective. Our study aims to bridge this knowledge gap by shedding light on the dimension of risk, as well as socio-economic and political factors, that constituted QOL for Japanese immigrants in Australia.

Drawing on the narratives of Japanese professional immigrants in Australia, this chapter will examine the multifaceted dimensions of QOL in Japan that motivated some of its citizens to leave the country. In the following sections, we first review the literature on QOL and emigration, followed by studies on Japanese immigration to Australia. By comparing and contrasting pre-2011 and post-2011 immigrants, we present our findings focusing on (1) the common dimensions of QOL that were perceived important by both pre-2011 and post-2011 Japanese immigrants and (2) the distinctive risks that specifically affected post-2011 Japanese immigrants' perceptions of QOL and their decision to leave Japan.

2 QOL and Migration: The Linkages Between QOL, Lifestyle, and Migration

QOL is a concept that broadly "covers all the aspects of human experience about the necessities of life" (Lin et al. 2013, p. 8). Since it entails multiple dimensions (Bowling and Windsor 2001, p. 55; Lin et al. 2013, p. 8), there is no single definition or measurement that can specifically fit across disciplines, cultures and time (Costanza et al. 2007, p. 268). The QOL Index (InterNations 2017), for example, uses five sub-indices to measure the total QOL level. These are leisure options, personal happiness, travel and transport, health and well-being, and safety and security. The Better-Life Index, however, adopts 11 sub-indices, including housing, work-life balance, education, and civic engagement (OECD 2017a) .

In our paper, we adopt Costanza's QOL definition, which is "the extent to which objective human needs are fulfilled in relation to personal or group perceptions of subjective well-being" (2007, p. 269). As seen in numerous empirical studies, lifestyle factors can influence QOL (Benson 2009, 2013; Benson and O'Reilly 2009; Ventegodt and Merrick 2003). Therefore, improving QOL is a major goal for many individuals, affecting their life choices (Schuessler and Fisher 1985). In the field of migration studies where labor migration had long been perceived as an economically-driven phenomenon, scholars began to address the importance of "lifestyle" or what "makes life worthwhile" (Alkire and Deneulin 2009, p. 25; Benson 2009, 2013; Benson and O'Reilly 2009). They argued that individuals, particularly educated skilled migrants from industrialized countries, are not driven purely by higher earnings, but rather choose their destination country based on a better quality of life and lifestyle. This so-called "lifestyle migration" is characterized by "non-economic" factors such as a "slower pace of life," "a simpler way of living," and "a closer connection with nature" (Benson and O'Reilly 2009). These scholars have argued that "lifestyle migrants" are distinguished from other immigrants from developing countries who primarily immigrate for better economic opportunities.

The existing empirical studies on Japanese citizens moving to Australia have been in line with this view (Mizukami 2007; Nagatomo 2008, 2015; Sato 2001), and have been framed in Benson and O'Reilly's framework of "lifestyle migration" (Benson and O'Reilly 2009). Khoo and her colleagues also found that 90% of Japanese skilled immigrants in Australia chose "lifestyles" as their primary reason for immigration (Khoo et al. 2007, p. 500). Other scholars (Mizukami 2007; Nagatomo 2008, 2015; Sato 2001) confirmed these findings, and echoed that lifestyle was the major driving force of migratory flows from Japan to Australia. More specifically, Mizukami (2007) also analyzed the nature of Japanese immigration to Australia as follows:

> To emigrate from Japan is not a matter of necessity or the pursuit of more attractive economic options, but something to do with an individual's desire for a change of social milieu in order to experience a renewed lifestyle … It can be said that Japanese immigrants to Australia give evidence of a desire to find a new, probably, transcending way of life (p. 175).

Based on his fieldwork and interviews with Japanese immigrants between 2005 and 2010, Nagatomo (2015) argued that growing tourism from Japan to Australia, as well as the culture of individualization in Japan, led young people to re-evaluate their working style and recognize the importance of leisure and lifestyle. He argued that the relaxation of immigration requirements and the growing leisure industry made it possible for many Japanese to immigrate to Australia to fulfil their desires. Many of his respondents were "searching for their own life fulfilment," and were "less concerned with future opportunities for their offspring" (Nagatomo 2015, p. 180). Furthermore, Nagatomo (2015) pointed out that gender inequality in the workplace and the family propelled many Japanese women to immigrate to Australia (pp. 116–119).

These studies greatly help us understand the patterns of Japanese immigration to Australia. Nevertheless, they do not fully capture the recent immigration patterns of Japanese professionals to Australia for three reasons. First, compared with the time these studies were conducted, Australia's immigration policies have become much more restrictive in terms of required skills, annual income, and English language proficiency[4] (Australian Government 2017, 2018; Azarias 2016). In response to the global financial crisis, the skilled occupations list for immigration was substantially revised in 2009 and subsequent years, enabling only highly educated and trained professionals to immigrate to Australia (Sprinks 2010: 4–5). This is in stark contrast with the era characterized by "the relaxation of entry requirements into Australia" in existing studies (Nagatomo 2015). Reflecting these major changes in immigration policies since 2009, recent Japanese immigrants are likely to have different characteristics and orientations compared with those who arrived in earlier eras.

Second, the impact of tourism as a precursor to Japanese immigration to Australia, which was highlighted in earlier studies (Mizukami 2007; Nagatomo 2008, 2015), seems unclear in recent years. Tourism from Japan to Australia has declined significantly in the last 20 years: the number of Japanese tourists almost halved from 814,000 in 1997 to 417,900 in 2016 because of economic downturns in Japan (Clarke 2018). With continuing deflation, Australia became a very expensive destination for Japanese tourists.

Third, after the Great East Japan Earthquake and the ensuing *tsunami* and Fukushima disasters in 2011, the mindsets of many Japanese seem to have changed: they now pay much more attention to diverse risks—radiation, earthquakes, volcano eruptions, and many other "country risks" that could affect their long-term QOL (Gordon 2014; Kimura 2016; Sternsdorff-Cisterna 2015). In recent years, a number of media reports have covered the increasing number of Japanese leaving the country (*Asahi Shimbun* 2012a, b; *Economist* 2016; *Nikkei Business* 2013; *Toyo Keizai* 2013, 2015), arguing that emigration has emerged as an option not only for the wealthy, but also for middle-class Japanese skilled workers and their families.

As seen in these developments in Japan and Australia's more restrictive immigration policies, we argue that the analytical framework of "lifestyle migration" does not fully capture the complex phenomenon of growing Japanese immigration to Australia. By comparing and contrasting the narratives of Japanese immigrants who arrived before and after 2011, we further argue that Japanese immigration to Australia has gone through a major shift from "migration as transnational leisure" (Nagatomo 2015) in the pre-2011 era to "migration as risk aversion" in the post-2011 era. Building on existing studies, this chapter demonstrates the prominent social, economic, and political dimensions of QOL in Japan that appear to have failed to meet the needs of Japanese professionals and to have thus propelled them to leave their country.

---

[4]For instance, the English proficiency requirement for permanent skilled migrants was raised from IELTS 4.5–6.0 (Australian Government 2010, 2018). The salary threshold for a skilled visa also increased from $34,075 in 2001 to $53,900 in 2013 (Azarias 2016), and the "minimum market salary rate" has also applied to the assessment of each application since March 2018 (Australian Government 2017).

3 Data and Methodology

This paper is based on the data obtained for a larger project that examines the migration of skilled migrants who left Japan for Hong Kong, Singapore, and Australia. For the selection of respondents, we used the OECD's definition of skilled workers, namely "human resources devoted to science and technology (HRST)" (OECD 1995). This definition has long been used in scholarly studies and reports of international organizations (e.g., Auriol and Sexton 2002; Zaletel 2006; Solimano 2004). It defines skilled workers as those who "successfully completed education at the third level in a science and technology field of study" or those who are "not formally qualified as above but employed in a science and technology occupation where the above qualifications are normally required" (OECD 1995, p. 18) . This is a very broad definition, since its scope of "science and technology" includes not only the hard sciences, but also the social sciences and humanities (OECD 1995, p. 21) .

A total of 32 respondents qualified as "skilled workers" by the OECD definition was identified through the authors' personal contacts and subsequent snowball sampling. Among them, 11 arrived in Australia before the Great East Japan Earthquake and Fukushima Disasters on March 11, 2011 (hereafter "pre-2011 immigrants"), and 19 arrived after these incidents ("post-2011 immigrants" hereafter). We also included two "prospective immigrants" in Tokyo who were still in the process of visa application at the time of the interview and arrived in 2018. The interviews with respondents who lived in Melbourne were conducted in person, while those with respondents in Adelaide, Cairns, Sydney, and Tokyo were conducted through Skype. Semi-structured questionnaires were used for all interviews, each of which lasted for 60–90 min. The interviews were recorded, transcribed and analyzed with NVivo11.

Regarding the profile of the respondents, 19 were male and 13 were female. About two-thirds (22) were in their 20s and 30s when they arrived in Australia; one-quarter (eight) were in their 40s. The two who were still in the visa application process made the decision to apply for skilled worker visas when they were in their late 30s. The average age at the time of arrival was 38.5 years. Except for three single respondents, all of them had a family with at least one child.

All respondents had a tertiary education: one person had a diploma from a technical school, and the rest had a bachelor's degree. In addition, one had a postgraduate diploma, three had a master's degree, and three had a Ph.D. degree. Four of them had a degree from overseas. The range of occupations was quite wide, including corporate executives, managers, engineers, consultants, entrepreneurs, medical workers, office clerks, technicians, and social workers. Reflecting Australia's recent immigration policies, the post-2011 respondents tended to have a higher education level and a higher occupational status than the pre-2011 respondents. As seen below, these two groups of respondents share some common perceptions of QOL, while distinctive differences also exist.

4 The Common Dimensions of QOL

The respondents who arrived before the 2011 disasters shared many of the characteristics addressed in the studies mentioned above (Mizukami 2007; Nagatomo 2008, 2015; Sato 2001). They discussed the importance of work-life balance, a relaxed lifestyle, leisure, and gender equality as the central drivers of immigration to Australia. In contrast, these factors were not very important for the immigration decision-making of most post-2011 respondents. After they arrived, however, they recognized them as important for their QOL in Australia. This section reviews the QOL dimensions—work-life balance, environment for raising children, and gender equality—that were positively addressed by both the pre-2011 and the post-2011 respondents.

4.1 Work-Life Balance

Corroborating the findings of Mizukami (2007) and Nagatomo (2008, 2015), work-life balance was the most prominent factor that the pre-2011 immigrants emphasized in their explanations of immigration decision-making. They felt work-life balance was a crucial component of good QOL. During the interviews, they often used QOL as a reference point to weigh the "push" and "pull" factors of immigration to Australia. While the domains of work and career were central in the respondents' narratives on QOL, they equally valued "non-work" (e.g., private, family, and community) life, and especially achieving a good work-life balance.

Indeed, work-life balance has become an increasingly pervasive concern, not only in Japan but also in many other countries (Fapohunda 2014, p. 72; Harrington and James 2006, p. 665). The issue is widely understood as "proper prioritizing between work (career and ambition) and lifestyle (health, pleasure, leisure, family, and spiritual development)" (Fapohunda 2014, p. 73). Finding a suitable balance between work and non-work life is hence considered critical for the well-being of workers and their families, and their overall quality of life (Greenhaus et al. 2003, pp. 510–531).

In our interviews, balancing work and life played a critical part in the pre-2011 respondents' search for a better QOL. They viewed living and working in Australia as an alternative way to meet their needs for a lifestyle that could help them to fulfil both work and non-work roles. Two of them recalled their work experience in Japan as follows:

> I was working for the office [travel agency] which was open until 9 pm. But as you know in Japan, even when a customer shows up at 9:02 or 9:03, you say, "How can I help you? Please come and sit down," and spend time with him/her. And after that, you do a bit more work, and then by the time you go home, you realize that it is the last train that you can catch. But then, the company's drinking party sometimes starts after that. I learned the difficulties of working life [in Japan].

> My job [in Japan] was quite demanding and I often felt like my life revolved around the company. So, I started looking for a job both in Japan and overseas that I could find more enjoyable and rewarding. Then I got a job in a X company in Melbourne, so I decided to move to Australia.

The phrases "work-life balance" and "work-family balance" were often cited when respondents provided the reasons why they decided to live and work in Australia. One remarked as follows:

> It's much easier to balance between work and family in Australia ... I value my relationship with children, so working here [in Australia] is great, because I can make choices about time that consider my children more easily [than in Japan]. I think that's the biggest advantage [of living and working in Australia] and the main reason why I decided to apply for a permanent resident visa.

Another respondent expressed similar views, pointing out "flexible work arrangements" and "shorter work hours" as the major characteristics of Australian workplaces. He explained how such aspects helped him to balance work and family life, observing that they were less attainable in Japan:

> I'm very happy at work [in Australia]. I mean, we have more flexible work arrangements [than in Japan]. So, I can spend more time with my family and kids. I sometimes take a day off just partaking in my kids' school events and engaging with other family commitments. And the great thing is that I don't have to feel guilty about it because that's what everyone else is doing here. In Japan, you know, it's a different story.

For many Japanese professionals, career success is often a critical component in determining life satisfaction and QOL. However, when they get an opportunity to live overseas and are exposed to a family-friendly work culture, they tend to re-examine their standard of QOL. Some of the respondents experienced a major change in their standard of QOL while living overseas, which led to their eventual decision to emigrate. One respondent used to work as an expatriate of a Japanese company in Australia and returned to Japan after his assignment. However, he later resigned from his company so that he could work in Australia again. He attributed his decision to the changes of his life stage, explaining that getting married and having a child changed his life priorities:

> [When working as a corporate expat in Australia] I was strongly feeling the difference in the standard of life [between Australia and Japan]. So, that was the biggest reason [for immigrating to Australia]. When I was working in Japan, I thought it was a very good experience. But I was working while having little sleep. On the other hand, I saw people really prioritizing their family [in Australia]. So, I wanted to have such a lifestyle. When I first came to Australia as an expat, I was single. So, work was a top priority for me. Then, I got married and had a child. Now, my priority is my family, although I like my work.

In contrast, only one post-2011 respondent mentioned lifestyle as the major reason for their immigration to Australia. Many of them were highly driven, hard-working professionals who were very used to working long hours and never questioned such a lifestyle. Because most of them had never worked in Australia or anywhere abroad, they could not imagine what the actual Australian lifestyle would be like. As we discuss in later sections, these post-2011 respondents cared more about diverse factors, such as environmental, economic, and socio-political risks. Nevertheless, after they arrived in Australia, they found that their lifestyle was slower and more relaxing, and they could enjoy more work-life balance. Those with a family were particularly pleased that they could spend more time with their family. Though the lifestyle was not their primary motivation for immigration to Australia, they were enjoying the side benefits of their decisions.

4.2 Environment for Raising Children

Earlier research found that Japanese immigrants' motivation for migration "was more self-directed, particularly in terms of searching for their own life fulfilment" and "less concerned with future opportunities for their offspring" (Nagatomo 2015, p. 181). However, all of the pre- and post-2011 respondents with a family considered the environment for raising children as a primary reason for immigrating to Australia.

The learning environment was particularly critical in the emigration decision-making of the respondents. In earlier studies, Mizukami (2007) discussed it in the framework of lifestyle migration, examining education as the core "push and pull" factor that affected Japanese immigrants (pp. 100–101). In Nagatomo's work (2008) on Japanese migrants in Australia, the respondents described the Japanese educational system as "stifling," "oppressive," and exam-oriented (pp. 230). In our study, some of the respondents shared this view. One stated:

> I was not very pleased with the Japanese education system. School teachers tell students to do the same thing and everyone follows. I didn't like that. I didn't want to do the same thing that other people were doing … [In Australia] school teachers do not deny students' individual characters. Students should be free. I think it is important that students can clearly express their opinions. If you express your opinions in Japan, you will be scolded.

Other respondents, however, were not too concerned with the strict education style in Japan, but were rather attracted by an Australian education that had more emphasis on critical thinking. All were highly educated individuals and thus some were expecting high educational achievements from their children as well. They wanted to develop their children's own views and critical thinking ability, and expose them to diverse ways of thinking.

The lack of a multicultural environment and an understanding of other cultures was another important factor for many respondents in deciding to leave Japan. One respondent and his Australian spouse decided to move to Australia not only because they perceived Japanese schools as being less diverse, but also because they believed that Japanese people did not have sufficient multicultural awareness. In fact, they were particularly concerned about the ways in which their daughter was treated in Japanese society. He recalled his experience in Japan:

> Our daughter is so called "a half" of Japanese and Australian. Then, people saw her in different ways, and couldn't accept her as a Japanese. I felt that way quite clearly … When we were riding a train, people said to our daughter, "Oh, so cute! You are like a doll. It is nice to be a *gaijin-san* [foreigner]." We were irritated by such comments. We didn't like it.

The majority of respondents also emphasized the significance of developing English proficiency and multicultural competency for their children's future careers. Given Japan's bleak economic future, they thought English language proficiency would open up more opportunities for their children. It should be noted, however, that their concerns were not just limited to language. Many parents emphasized the importance of diversity and a multicultural environment. Given their own work experience, the respondents perceived intercultural understanding, literacy, and competency as crucial assets for developing careers in the global labor market. In their view, an Australian education would help their children acquire these skills and competencies, enabling them to work with people with different cultural backgrounds very easily. They had pessimistic views on the long-term prospects of the Japanese economy and expected that domestic employment opportunities would decline in the future. In response to their perception of risk in the Japanese economy and labor market, they wanted to provide their children with better opportunities and more future options through an Australian multilingual and multicultural education, believing that it would lead them to a better quality of life.

Last but not least, physical safety played a major role in the respondents' assessment of potential destination countries. One person emphasized the importance of safety as follows:

> The biggest reason why we came to Australia was children. We made the decision when they were about to enter a primary school … So, we cared about safety. We wouldn't have considered [immigrating to] a country with a problem of safety. Safety was important.

Another respondent, who was a mother of two elementary-school children at the time of the interview, shared a similar view, commenting that:

> Safety and public order were the most important factors when choosing our destination country. I lived in the United States when I was a university student and didn't really feel safe walking even during the day. I also didn't feel that it would be a safe environment for my children. But now I feel safe [in Australia] and I think it's a good and safe place for my children, too.

Since Japan is known for its low crime rate, safety was not their reason for immigration to Australia. However, when they were finalizing their decision on their destination country, low crime rates and lack of gun problems were major factors that they took into consideration. Some respondents chose Australia partly because of the strict gun-control systems and the image of its relatively low crime rates compared with other countries. Another "safety" reason was the absence of nuclear power plants in Australia, which we discuss in a later section.

4.3 Gender Equality

Recent research demonstrates the benefits gender equality has for QOL (Holter et al. 2009; Holter 2014; Warat et al. 2017). Based on data from Europe and the United States, Holter (2014) suggested that both women's and men's QOL is higher in countries or states with a relatively high degree of gender equality, compared with those with a low degree of gender equality. In the context of Poland and Norway, Warat et al (2017) also highlighted gender equality as an important factor in QOL, arguing that achieving gender equality allows both women and men to achieve full and equal participation in many spheres of life. The OECD (2014) confirmed these findings, reporting that living in a country guaranteeing equal rights and opportunities to women and men increases people's happiness. The OECD Development Centre's study (2014) further suggested that gender-based discrimination in social institutions impedes well-being beyond GDP, demonstrating that current gender-based discrimination in social institutions fuels a decline of 4.4% in the world average level of life satisfaction.

In the case of Japan, gender inequality is pervasive and has long been known as the worst among industrialized countries. The country ranked 114th among 144 countries in the World Economic Forum's Global Gender Gap Index (WEF 2017) . Japan is also at the bottom of the Glass Ceiling Index compiled by *The Economist* magazine in 2017, ranking 28th among 29 countries (*Economist* 2017). Moreover, the country has the third-largest gender wage gap in the OECD countries: women earn 24.5% (a percentage of male median wage) less than men as of 2017, after Ethiopia (28.3%) and South Korea (34.6%) (OECD 2018a). Such a serious degree of gender inequality has had an impact on Japanese women's QOL as well as their eventual decision for emigration. In migration studies, Japan's gender inequality has been discussed as the prominent factor contributing to the outflow of Japanese women. Many scholars revealed that Japanese women's emigration is a response to limited career opportunities, a resistance to gendered expectations, and an exploration of different cultures (Kelsky 2001; Nagatomo 2008, 2015; Sato 2001; Toyota 2006).

In the interviews, some of the female respondents shared stories about their experiences of gender stereotyping and discrimination in Japan and how this had affected their career choices and successes. One female physician pointed out the different treatment of female employees:

> I was thinking of developing my career as a specialist. But during the job interview, I was told, "We don't want women" – straightforwardly. "Women get pregnant and stop working for a while during your career," he said. [Was the interviewer a male?] Yes. He was biased and too blunt.

This respondent eventually landed a job at a different hospital as a specialist and continues working while raising a child. However, this kind of inappropriate comment is relatively common in Japan, since many employers assume that female candidates resign after they get married and/or have children. While 46.9% of Japanese women indeed leave work after having their first child (IPPS 2017), it is widely known that this is due to a harsh, family-unfriendly work environment, such as long work hours.

This has led some companies to introduce new systems or improve work conditions to retain female professionals (Cabinet Office 2014). Nevertheless, some traditional employers still emphasize the importance of working long hours, believing that sacrificing one's time for work represents "commitment, endurance, and toughness" (Nemoto 2012, p. 514). Therefore, they perceive women as not being as committed to their work as men and thus are reluctant to offer employment and promotion opportunities (Nemoto 2012).

Although the Equal Employment Opportunity Law (EEOL) was enacted in 1986, many large Japanese companies introduced a particular hiring system to limit women to a "clerical, non-career track" as opposed to a managerial career track (Senda 2015, pp. 51–82). Under this system, more women are recruited for non-career-track positions with lower wages and limited career opportunities (Senda 2015, pp. 51–82). Partly because of this, Japan has the lowest proportion of female managers in its workforce, whereas Australia has the highest proportion among 31 OECD countries (OECD 2018b). Our interview data affirmed the impact of gender inequality in Japan on Japanese women's careers, as suggested in previous studies, particularly on Japanese professional women who immigrated to Australia as a way to improve their QOL. One female respondent stated:

> I came here [to Australia] to find a better-paying job because my pay in Japan was very low … I still believe that gender and age are the biggest obstacles when it comes to [women] working in Japan.

In this study, the female respondents perceived gender equality as a critical component determining their career and life satisfaction, and thus their overall QOL. They expressed their desire for living in a gender-equal society and being empowered. In the interviews, they highlighted Japan as a "highly gendered society." Notably, gender inequality was directly and indirectly attributable to their decision to leave the country. Living and working in Australia was an alternative option for them to fulfil their desire for empowerment and to achieve a better QOL.

5 Distinctive Differences in the Pre- and Post-2011 Japanese Immigrants

As discussed so far, all the respondents appreciated some common lifestyle-related QOL factors in Australia—a better work-life balance, quality education and a safe environment for children, and greater gender equality. Nevertheless, there were stark differences in that these factors were significant enough to motivate all pre-2011 respondents to immigrate to Australia, whereas this was not the case for most post-2011 respondents. The factors that affected post-2011 respondents had more to do with a wide range of risks that could have significant implications for long-term QOL for themselves and their children. This difference suggests the possibility of a major shift in the perception of and concerns for QOL by the Japanese public before and after the Great East Japan Earthquake and Fukushima disasters. This section provides an overview of QOL-related factors—environmental, socio-economic, and political—that compelled post-2011 respondents to leave Japan for Australia.

5.1 Disaster Risks and QOL

The most critical dimension of QOL that emerged prominently in our post-2011 respondents' narratives of their emigration decision was environmental risks. Overall, 81% of the post-2011 respondents raised potential natural and nuclear disasters as one of the reasons for leaving Japan. Even though none of them had lived in Tohoku, they were all concerned with the impacts of the disasters and future risks.

For the first few years after the Fukushima disasters, the lack of information and "fear of the unknown" caused anxieties for the post-2011 respondents. Many were worried about the risk of being exposed to radiation, since the Japanese government had strictly controlled the public's access to information on radiation and what was happening in Fukushima. One respondent even bought a Geiger counter because his wife was extremely worried about the impact of radiation for their son. Most respondents with school-aged children were particularly concerned because they could not get information on the origins of ingredients in school meals or restaurant food. The majority of respondents commented that they were anxious trying to secure safe food for themselves and their families. One typical example was the following:

> Since 3.11 [disasters], we have had to pay close attention to our food every day because of radiation. Even now [in 2017], we only select products from the western part of Japan in supermarkets. This was added to our life as an additional task. This issue, which we had never thought of before, emerged as our everyday concern. This has become a major stress.

While acute concerns about radiation exposure for children gradually dissipated over time, people were still emigrating even six to seven years after the Fukushima disasters because of the environmental risks. This is partly because of the widely-shared view among scientists and policy makers that Japan entered the geologically active era after 2011. Soon after the Great East Japan Earthquake, the Japanese government as well as the scientific community disclosed the reality of Japan's geological risks. Based on the researchers' analyses, the Japanese government issued an official prediction that the chance of M8–M9 class earthquakes striking central and/or western Japan within 30 years is 60–70% (Headquarters for Earthquake Research Promotion 2013, p. 6), with an estimate of up to 323,000 casualties (Cabinet Office 2012, p. 21). They also disclosed the consensus of the researchers that the chance of M7–8 class earthquakes hitting the Tokyo area within 30 years is 70%, with estimated casualties of 23,000 (Central Disaster Prevention Council 2013, pp. 8–13). While these major earthquakes have not occurred yet, there have been nine M7+ earthquakes in Japan since the Great East Japan Earthquake in 2011 (Ministry of Education, Culture, Sports, Science and Technology 2018). Even smaller earthquakes that hit Nagano (M6.7) in 2011, Kumamoto (M6.3 and M7) in 2016, Tottori (M6.6) in 2016, and Osaka (M5.5) in 2018 caused major damage.

Japan also has a number of active volcanoes, including Mount Fuji, which is relatively close to Tokyo. The government acknowledged that Mount Fuji and other volcanoes are likely to erupt at some point in the future, which would significantly affect the infrastructure of the Greater Metropolitan Tokyo area, and policymakers began to take various measures to minimize the damage (Cabinet Office 2005).

Many post-2011 respondents felt that the natural disasters predicted could cause similar nuclear disasters as occurred in Fukushima, given the fact that the Japanese government had maintained its nuclear energy policy. While the majority of Japanese citizens are against the operation of nuclear power plants (*Mainichi Shimbun* 2017; *Nihon Keizai Shimbun* 2016), the Japanese government re-started the operation of some of them. Many respondents expressed their grave concerns and raised it as one of the reasons for immigrating to Australia where there are no nuclear power plants. Two respondents who still lived in Tokyo at the time of the interview expressed their anxieties as follows:

> Japan has nuclear power plants, while being an "earthquake country." We also have *tsunami*. And there is no evacuation plan or measures [against possible future disasters]. But the government is still proceeding with the re-activation of nuclear power plants. I am very worried about this.

> Japan is said to have entered the geologically active period… In fact, there was a big earthquake in Kyushu – in Kumamoto recently [2016]. Even then, the Sendai nuclear power plant [in the prefecture next to Kumamoto] is still operating … The issue of earthquakes is always in my mind … Australia also has earthquakes, and it is impossible to completely escape from them. But at least we want to be away from the combination of earthquakes and nuclear power plants. When we stayed in Australia for two months [in early 2017], we loved it – we were released from anxieties and were stress-free.

The natural disasters predicted, combined with potential nuclear disasters, was causing stress for the post-2011 respondents, lowering their QOL. They knew that if these concerns materialized, their QOL would be acutely jeopardized, much more so than in the case of the Great East Japan Earthquake and Fukushima disasters. These thoughts were substantial enough to compel them to leave Japan for a place where nuclear power plants did not exist.

5.2 *Long-Term Economic Risks and QOL*

A long-term concern about the bleak prospects of the Japanese economy was another point of distinctive difference between the pre-2011 and post-2011 respondents' narratives. While none of the pre-2011 respondents mentioned the country's long-term economic prospect in their interview, 87.5% of the post-2011 respondents cited it as one of the primary reasons why they left Japan. Post-2011 respondents believed that the acceleration of population aging and the lack of immigration to Japan would inevitably shrink the country's labor force, as well as its consumer market. In addition to Japan's overall economic future, they expressed acute concern about their own financial viability after retirement—namely, the sustainability of the Japanese old-age pension system, under which current workers' contributions are used to pay the benefits of current retirees. With a shrinking workforce, they began to seriously doubt that they would receive the benefits that they are now entitled to, as the birth rate is expected to fall further in the coming decades.

The fact that the Japanese government has not taken any effective measures against accelerating population decline exacerbated their pessimism about the country's future. Some respondents especially expressed their grave concerns about the fact that Japan has the world's highest national debt of 1,049.37 trillion yen (US$9.5 trillion), which is 234% of its GDP (OECD 2017b). One respondent even mentioned the possibility of a more extreme case:

> Because the situation of Japanese economy is deteriorating, if we enter wars with the US against North Korea or if another huge earthquake or nuclear disaster occurs again, things like deposit blockade[5] might happen, I think, though this is the worst-case scenario. In fact, the government did a deposit blockade at the end of WWII and all Japanese became penniless. So, nobody can say that it will never happen again.

Family status has made a distinct difference between pre-2011 respondents and post-2011 respondents as well. The vast majority of pre-2011 respondents were single or married without children when they arrived in Australia. Therefore, none of them mentioned any issues with children as a reason for leaving Japan. Most post-2011 respondents, however, were married and had at least one child at the time of immigrating to Australia. For them, future economic concerns for their children constituted a major component of their emigration decision. Given the accelerating population aging and shrinking market, post-2011 respondents felt that domestic employment opportunities for their children would be much more limited in the future. By immigrating to Australia, they hoped to provide their children with better opportunities and more future options through a multilingual and multicultural education, believing that it would help lead them to a better quality of life. For them, immigration was a hedging strategy to secure financial stability, which is one of the most important foundations for their long-term QOL and well-being.

5.3 The Potential Long-Term Impacts of Socio-political Factors on QOL

Another distinct difference between the pre-2011 and post-2011 respondents was that the latter were much more politically aware and critical of the current government's policies on nuclear energy, international security, old-age pensions, information control, and so on. The vast majority were quite outspoken about their distrust of the government and expressed their dissatisfaction with the political developments that could significantly affect their lives. This was in sharp contrast to the pre-2011

[5]The emergency measure that governments could take to stop their citizens' withdrawal of money from their bank accounts. It happened in Japan at the end of WWII, as well as in other countries such as Greece and Malta in more recent years.

respondents who did not mention any political factors or the impacts of such factors on their QOL.

The majority of the post-2011 respondents were particularly concerned with the risks that these policies might entail and their potential impact on their own and their family's QOL. The specific concerns included new legislation such as Security Laws, the State Secrecy Law, and the Constitutional reform that will enable the government to enter into war. Some respondents thought these changes could lead Japan to engage in military conflict with neighboring countries in the near future. The media reports on the possible introduction of a conscription system to expand Japan's self-defense forces also caused anxieties for parents with young sons. These political developments have indeed affected some respondents' decision to leave Japan. The following two comments epitomize the views of these parents:

> Our decision to move to Australia was initially triggered by fear of radiation. But the second biggest reason was the current LDP policies such as Collective Self-Defense, Security Bills and the Constitutional Reform … We support the current Constitution [which prohibits Japan from exercising military power other than for self-defense purposes]. We don't want wars. What worried us most was the introduction of conscription system, and our son might be taken away from us.

> As you know, the Conspiracy Law was just passed. Before that, the State Secrecy Law and Security Laws were passed. There have been many developments. So, it seems that Japan is probably on the path to becoming a country that will be able to start wars. When a war happens, however, deposit blockade could happen and there will be many things that Japanese citizens will have to persevere. That is not the direction that we want … We have been feeling great stress. Our anxiety has been growing.

As elucidated in these narratives, some respondents believed that peace constitutes a solid foundation of high QOL, since military conflicts with other countries could disrupt people's lives and damage their socio-economic, financial, and physical well-being tremendously. Having learned that the current Japanese government was eager to revise the Constitution and might expand its self-defense force into a greater national military force, the post-2011 respondents decided to avoid any potential physical and socio-economic risks, as well as the environmental risks and natural disasters as discussed above. The move of the post-2011 respondents from Japan to Australia was clearly a risk-averse migration.

5.4 *Life Stage and QOL*

As discussed above, sensitivity to risks was a major analytical axis that demarcated pre-2011 and post-2011 Japanese immigrants in Australia. At the same time, such a difference partly derived from their family status and also reflected different perceptions of QOL in different life stages. While young single immigrants were driven mostly by career development and gender equality, those with a family were driven more by factors related to their children, including no nuclear power plants, a safe environment, better education, and better economic opportunities for children, as major incentives for immigration.

Interestingly, some of the immigrants—both pre-2011 and post-2011—felt that QOL in Japan for retirees was higher in Japan than in Australia. For this reason, 18.8% of them were considering the option of returning to Japan. In their view, corporate life in Japan required long working hours and lacked work-life balance, but retirement life seemed nice and relaxing. Partly because of the lower living cost in Japan and the culture of respecting the elderly, they felt that QOL for retirees in Japan would be better than in Australia. They also believed that retirees in Japan would have many opportunities to enjoy excellent food and rich cultural activities. One respondent commented:

> I don't think I will stay here for good. Because if I get old, I would probably feel that I want to see my old friends. I think it will be much more fun to live in Japan for older people. Japan has more events than in Australia. Regional areas have their own unique cultural events. Even if you make a short trip, it is fun. Here, wherever you go, you see the same things. Once I retire, I will return to Japan.

Even those who were planning to settle in Australia for good felt that they would like to spend some time in Japan every year for the same reason. This indicates that the perception of QOL is not static or monolithic, but rather has distinctive life stage components. The QOL perceptions could shift as the respondents' lifestyles and priorities change over time.

6 Conclusion

This chapter presented a critical examination of QOL in Japan from the perspective of its citizens who have left the country. By analyzing the narratives of 32 professionals who left Japan before and after the Great East Japan Earthquake and Fukushima disasters, it examined the ways in which they perceived their QOL in Japan and Australia, as well as diverse potential factors that might affect their long-term QOL. Our study provided evidence for the complex linkages between the continuing skilled migration from Japan to Australia and the issues affecting the immigrants' perceptions of QOL. It became apparent through the analysis of their narratives that they constantly used their QOL in Japan as a reference point to explain the reasons why they had decided to immigrate to Australia.

The analysis of our qualitative interview data demonstrated that there were distinct differences in the motivations for immigration and the ways in which QOL was perceived between the pre-2011 and the post-2011 respondents. The pre-2011 respondents, who were mostly unmarried and/or had no children, highlighted the importance of work-life balance and gender equality. The general orientation of the pre-2011 respondents was in line with the existing literature, which had long framed Japanese immigration to Australia as "lifestyle migration" (Mizukami 2007; Sato 2001) and "migration as transnational leisure" (Nagatomo 2015). While the post-

2011 respondents appreciated these QOL components, these were not necessarily the main factors that compelled them to leave their country.

Those who chose to leave Japan after the Great East Japan Earthquake and the ensuing Fukushima disasters in 2011 were not seeking a more relaxed or leisurely lifestyle. Rather, they were driven by risk aversion, aiming to protect themselves and their families from both short-term and long-term risks that entailed major QOL implications. The majority of the post-2011 respondents left Japan because of the environmental risks and long-term economic and socio-political factors that could severely undermine the short-term and long-term QOL and well-being for themselves and their family members.

This research offered a pioneering analysis of the linkages of migration, risks, and QOL. It also highlighted the importance of the temporal dimension of QOL, particularly individuals' changing perceptions of QOL as well as their long-term assessment of future risks and the potential impact of these risks on QOL. The authors acknowledge the limitation of this study due to its relatively small scope. Further case studies on other destination countries, incorporating the role of life stages, would be needed to clarify the complexities of migration decision-making and the role of QOL across time and space. The sudden increase (43.9%) of Japanese permanent immigrants in Germany, for instance, might reflect the German government's decision in 2011 to shut down all nuclear power plants by 2022 (MOFA 2018; BBC 2011). Anecdotal evidence also suggests that low-tax countries such as Singapore and New Zealand have been attracting wealthy Japanese individuals who, after 2011, wished to protect their assets from future risks (*Asahi Shimbun* 2012b; *Diamond Online* 2011). More comparative country case studies are needed to fully understand the broader picture of post-3.11 emigration from Japan and the emigrants' changing QOL perceptions.

Acknowledgement Dr. Nana Oishi is Associate Professor in Japanese Studies at the Asia Institute, The University of Melbourne. Dr. Iori Hamada is Lecturer in Japanese Studies at Monash University. We would like to thank the Editorial Board and three anonymous reviewers for their valuable advice and insightful comments on an earlier draft of this paper.

References

Alkire, S., & Deneulin, S. (2009). The human development and capability approach. In S. Deneulin & L. Shahani (Eds.), *An introduction to the human development and capability approach: Freedom and agency* (pp. 22–48). London: Earthscan/IDRC.

Asahi Shimbun. (2012a). *Protection from radiation: Exiting Japan* (in Japanese). January 1. Retrieved April 10, 2018, from Kikuzo Database.

Asahi Shimbun. (2012b, January 4). *Fear for financial collapse: Drifting money and people* (in Japanese). Retrieved April 10, 2018 from Kikuzo Database.

Auriol, L. & Sexton, J. (2002). Human resources in science and technology: Measurement issues and international mobility. In OECD (ed), *International mobility of the highly skilled*. Paris: OECD.

Australian Government. (2010, *February 8*). *Skilled occupation lists (SOL) in effect.* Retrieved May 28, 2018, from http://www.homeaffairs.gov.au/Migrationagents/Documents/sol-twoabc.pdf.

Australian Government. (2017). *Fact sheet two: Reforms to Australia's permanent skilled migration program*. Retrieved June 4, 2018, from http://www.homeaffairs.gov.au/WorkinginAustralia/Documents/reforms-australia-permanent-employer-sponsored-migration-programme.pdf.
Australian Government. (2018). *How can I prove I have competent English?* Retrieved June 4, 2018, from http://www.homeaffairs.gov.au/lega/lega/form/immi-faqs/how-can-i-prove-i-have-competent-english.
Azarias, J. (2016). *Review of the Temporary Skilled Migration Income Threshold (TSMIT)*. Retrieved June 4, 2018, from http://www.homeaffairs.gov.au/ReportsandPublications/Documents/reviews-and-inquiries/tsmit_review_report.pdf.
Benson, M. (2009). A desire for difference: British lifestyle migration to southwest France. In M. Benson & K. O'Reilly (Eds.), *Lifestyle Migration: Expectations, aspirations and experiences* (pp. 121–136). Farnham: Ashgate.
Benson, M. (2013). Living the 'real' dream in La France Profonde? Lifestyle migration, social distinction, and the authenticities of everyday life. *Anthropological Quarterly, 86*(2), 501–525.
Benson, M., & O'Reilly, K. (2009). Migration and the search for a better way of life: A critical exploration of lifestyle migration. *Sociological Review, 57*(4), 608–625.
Bowling, A., & Windsor, J. (2001). Towards the good life: A population survey of dimensions of quality of life. *Journal of Happiness Studies, 2*(1), 55–82.
British Broadcasting Corporation (BBC). (2011). *Germany: Nuclear power plants to close by 2022*. Retrieved August 11, 2018, from http://www.bbc.com/news/world-europe-13592208.
Cabinet Office. (2005). *Fujisan kazan kouiki bōsaitaisaku kentōkai hōkokusho* (*The council report on the cross jurisdictional measures against Mt. Fuji volcano*). Retrieved May 15, 2017, from http://www.bousai.go.jp/kazan/fujisan/w_g/kentou/houkoku/pdf/siryo0.pdf.
Cabinet Office. (2012). *Nankai torafu kyodai jishin no higaisōteinitsuite* (*The report on the potential damage of Nankai megathrust earthquakes*). Retrieved August 15, 2017, from http://www.bousai.go.jp/jishin/nankai/taisaku/pdf/20120829_higai.pdf.
Cabinet Office. (2014). *Danjo kyōdō sankaku hakusho heisei 26 nendo ban* (*2014 whitepaper on making a gender-equal society*). Retrieved May 28, 2018, from http://www.gender.go.jp/about_danjo/whitepaper/h26/zentai/index.html.
Central Disaster Prevention Council. (2013). *Shutochokkagatajishin no Higaisōteito taisaku* (*The damage estimates for Tokyo inland earthquake and the measures: Final report*). Cabinet Office, Government of Japan.
Clarke, C. (2018, January 16). Australia and Japan reverse tourism relationship. *ABC News Analyses*. Retrieved June 5, 2018, from http://www.abc.net.au/news/2018-01-16/australia-and-japan-reverse-tourism-relationship/9332518.
Costanza, R., et al. (2007). Quality of life: An approach integrating opportunities, human needs, and subjective well-being. *Ecological Economics, 61*(2–3), 267–276.
Diamond Online. (2011). *Financial crisis, radiation contamination—The wealthy class started leaving Japan fearing "Japan risk"* (in Japanese). Retrieved May 15, 2017, from http://diamond.jp/articles/-/14258.
The Economist. (2017). *The glass-ceiling index*. Retrieved November 19, 2017, from http://infographics.economist.com/2017/glass-ceiling/.
The Economist. (2000, June 29). *Quality of life*. Retrieved May 19, 2018, from http://www.economist.com/node/3276.
Economist (Japanese weekly). (2016, May 24). *Sneaky tax evasion*. Special Issue.
Fapohunda, T. M. (2014). An exploration of the effects of work life balance on productivity. *Journal of Human Resources Management and Labor Studies, 2*(2), 71–89.
Gordon, A. (2014). *A modern history of Japan: From Tokugawa times to the present* (3rd ed.). Oxford University Press: New York.
Greenhaus, J. H., Collins, K. M., & Shaw, J. D. (2003). Work-family balance and quality of life. *Journal of Vocational Behavior, 63*(3), 510–531.
Harrington, B., & James, J. B. (2006). The standards of excellence in work-life integration: From changing policies to changing organizations. In M. Pitt-Catsouphes, E. E. Kossek, & S.

Sweet (Eds.), *The work and family handbook: Multi-disciplinary perspectives and approaches* (pp. 665–683). Mahwah, NJ: Lawrece Erlbaum.

Headquarters for Earthquake Research Promotion (HERP). (2013). *Long-term assessment of Nankai trough earthquake activities* (in Japanese). Retrieved May 28, 2018, from http://www.jishin.go.jp/main/chousa/13may_nankai/nankai2_shubun.pdf.

Holter, Ø. G. (2014). 'What's in it for men': Old question, new data. *Men and Masculinities, 17*(5), 515–548.

Holter, Ø. G., Svare, H., & Egeland, C. (2009). *Gender equality and quality of life—A norwegian perspective*. Oslo, Norway: NIKK (Nordic Gender Institute).

InterNations. (2017). *Quality of life index, expat insider 2017: Looking at the world through expat eyes*. Retrieved, May 19, 2018, from http://www.internations.org/expat-insider/2017/quality-of-life-index-39185.

Kelsky, K. (2001). *Women on the verge: Japanese women, Western dreams*. Durham: Duke University Press.

Khoo, S. E., McDonald, P., Voigt-Graf, C., & Hugo, G. (2007). A global labour market: Factors motivating the sponsorship and temporary migration of skilled workers to Australia. *International Migration Review, 41*(2), 480–510.

Kimura, A. H. (2016). *Radiation brain moms and citizen scientists: The gender politics of food contamination after Fukushima*. Durham, NC: Duke University Press.

Lin, X., Lin, I., & Fan, S. (2013). Methodological issues in measuring health-related quality of life. *Tzu Chi Medical Journal, 25*(1), 8–12.

Mainichi Shimbun. (2017, March 13). *Opinion poll: 55% oppose and 26% support the reactivation of nuclear power plants*.

Ministry of Education, Culture, Sports, Science and Technology. (2018). The evaluation of main geological activities. *The Headquarters for Earthquake Research Promotion*. Retrieved August 11, 2018 from http://www.jishin.go.jp/evaluation/seismicity_annual/major_act_2011/.

Ministry of Foreign Affairs (MOFA). (2002). *Annual report of statistics on Japanese nationals overseas*. Retrieved May 21, 2018, from http://www.mofa.go.jp/mofaj/toko/tokei/hojin/12/pdfs/WebBrowse.pdf.

Ministry of Foreign Affairs (MOFA). (2012). *Annual report of statistics on Japanese nationals overseas*. Retrieved May 21, 2018, from http://www.mofa.go.jp/mofaj/toko/tokei/hojin/01/index.html.

Ministry of Foreign Affairs (MOFA). (2018). *Annual report of statistics on Japanese nationals overseas*. Retrieved August 20, 2018, from https://www.mofa.go.jp/mofaj/files/000368753.pdf.

Mizukami, T. (2007). *The sojourner community: Japanese migration and residency in Australia*. Koninklijke Brill: Leiden.

Nagatomo, J. (2008). Globalisation, tourism development, and Japanese lifestyle migration to Australia. In D. M. Nault (Ed.), *Development in Asia: Interdisciplinary, post-neoliberal, and transnational perspectives* (pp. 215–236). Boca Raton: Brown Walker Press.

Nagatomo, J. (2015). *Migration as transnational leisure: The Japanese lifestyle migrants in Australia*. Leiden: Brill.

National Institute of Population and Population and Social Security Research (IPPS). (2017). *Marriage and childbirth in Japan today: The fifteenth Japanese national fertility Survey, 2015*. Survey Series No. 35.

Nemoto, K. (2012). Long Working hours and the corporate gender divide in Japan. *Gender, Work and Organization, 20*(5), 512–527.

Nihon Keizai Shimbun. (2016, February 29). *Opinion poll: 60% oppose to the reactivation of nuclear power plants*.

Nikkei Business. (2013, July 8). *Tokushu: Jinzai tōhi* (Special issue: Flight of human resources). Volume No. 1698.

Organization for Economic Cooperation and Development (OECD). (1995). *The measurement of scientific and technological activities. Manual on the measurement of human resources devoted to*

S&T. Canberra Manual. Retrieved on May 28, 2018, from http://www.oecd-ilibrary.org/science-and-technology/the-measurement-of-scientific-and-technical-activities_9789264063525-en.

Organization for Economic Cooperation and Development (OECD) Development Centre. (2014), *Social institutions and gender index (SIGI) 2014 synthesis report*. Retrieved November 19, 2017, from http://www.oecd.org/dev/development-gender/BrochureSIGI2015-web.pdf.

Organization for Economic Cooperation and Development (OECD). (2017a). *Better life index*. Retrieved May 21, 2018, from http://www.oecdbetterlifeindex.org/#/11111111111.

Organization for Economic Cooperation and Development (OECD). (2017b). *General government debt (Indicator)*. Retrieved June 25, 2017, from http://dx.doi.org/10.1787/a0528cc2-en.

Organization for Economic Cooperation and Development (OECD). (2018a). *Gender wage gap (indicator)*. Retrieved 25 August 2018, from https://data.oecd.org/earnwage/gender-wage-gap.htm.

Organization for Economic Cooperation and Development (OECD). (2018b). *Employment: share of employed who are managers, by sex*. Retrieved May 21, 2018, from http://stats.oecd.org/index.aspx?queryid=54752.

Sato, M. (2001). *Farewell to Japan*. Melbourne: Trans Pacific Press.

Schuessler, K. F., & Fisher, G. A. (1985). Quality of life research and sociology. *Annual Review of Sociology, 11*, 129–149.

Senda, Y. (2015). *Child bearing and careers of Japanese women born in the 1960s*. Tokyo: Springer.

Solimano, A. (2004). *International mobility of the highly skilled: The case between Europe and Latin America*. Paris: OECD.

Sprinks, H. (2010, October 29). *Australia's migration program*. Parliament of Australia, Social Policy Section. Parliamentary Library, Information Analysis Advice.

Sternsdorff-Cisterna, N. (2015). Food after Fukushima: Risk and scientific citizenship in Japan. *American Anthropologist, 117*(3), 455–467.

Toyo Keizai. (2013, February 4). *Kaigai iju to toshi: datsu nippon toiu sentaku* (Overseas migration and investment: the option of leaving Japan). Special Issue.

Toyo Keizai. (2015, February 21). *Honkide kangaeru kaigai iju & shisan unyo* (Serious thoughts on overseas migration and asset management). Special Issue.

Toyota, M. (2006). Ageing and transnational householding: Japanese retirees in Southeast Asia. *International Development Planning Review, 28*(4), 515–531.

United Nations Development Programme (UNDP). (2016). *Human development report 2016*. New York: United Nations.

Ventegodt, S., & Merrick, J. (2003). Lifestyle, quality of life, and health. *The Scientific World Journal, 3*, 811–825.

Warat, M., Krazaklewska, E., Ratecha, A., & Slany, K. (Eds.). (2017). *Gender equality and quality of life: Perspectives from Poland and Norway*. Frankfurt am Main: Peter Lang GmbH.

World Economic Forum. (WEF). (2017). *The global gender gap report 2017*. Retrieved May 19, 2018, from www3.weforum.org/docs/WEF_GGGR_2017.pdf.

Zaletel, P. (2006). Competing for the highly skilled migrants: Implications for the EU common approach on temporary economic migration. *European Law Journal, 12*(5), 613–635.

Nana Oishi is Associate Professor in Japanese Studies at the University of Melbourne, Australia. She completed her Ph.D. degree in Sociology at Harvard University as a Fulbright Scholar. Prior to her current position, she was Professor of Sociology at Sophia University in Tokyo and Policy Analyst at the International Labour Organization (ILO) in Geneva. She has been conducting research on migration issues for the last 25 years, and her articles have appeared in *American Behavioral Scientist, Journal of Ethnic and Migration Studies*, and *Policy Studies*. Her recent research projects address skilled migration, diversity and inclusion in Japan and Australia (nana.oishi@unimelb.edu.au).

Iori Hamada is Lecturer in Japanese Studies at Monash University, Australia. Her research brings transnational perspectives to bear on investigating key areas in contemporary culture and society: migration, social inclusion, gender ideologies, food cultures, higher education and the work-family nexus. Her work has appeared in leading international publishers, including Springer, Oxford University Press, Routledge and Bloomsbury Academic (Iori.Hamada@monash.edu).

Index

M.-C. Tsai and N. Iwai (eds.), *Quality of Life in Japan*, Quality of Life in Asia 13,
https://doi.org/10.1007/978-981-13-8910-8